Improving Healthcare
through
Built Environment Infrastructure

Improving Healthcare
through
Built Environment Infrastructure

Edited by

Mike Kagioglou
Programme Director,
Health and Care Infrastructures Research and
Innovation Centre,
Head of the School of the Built Environment,
University of Salford, UK

Patricia Tzortzopoulos
Academic Fellow
School of the Built Environment
University of Salford, UK

WILEY-BLACKWELL

A John Wiley & Sons, Ltd., Publication

This edition first published 2010
© 2010 Blackwell Publishing Ltd

Blackwell Publishing was acquired by John Wiley & Sons in February 2007. Blackwell's publishing programme has been merged with Wiley's global Scientific, Technical, and Medical business to form Wiley-Blackwell.

Registered office
John Wiley & Sons Ltd, The Atrium, Southern Gate, Chichester, West Sussex, PO19 8SQ, United Kingdom

Editorial offices
9600 Garsington Road, Oxford, OX4 2DQ, United Kingdom
350 Main Street, Malden, MA 02148-5020, USA

For details of our global editorial offices, for customer services and for information about how to apply for permission to reuse the copyright material in this book please see our website at www.wiley .com/wiley-blackwell.

The right of the author to be identified as the author of this work has been asserted in accordance with the UK Copyright, Designs and Patents Act 1988.

Wiley also publishes its books in a variety of electronic formats. Some content that appears in print may not be available in electronic books.

Designations used by companies to distinguish their products are often claimed as trademarks. All brand names and product names used in this book are trade names, service marks, trademarks or registered trademarks of their respective owners. The publisher is not associated with any product or vendor mentioned in this book. This publication is designed to provide accurate and authoritative information in regard to the subject matter covered. It is sold on the understanding that the publisher is not engaged in rendering professional services. If professional advice or other expert assistance is required, the services of a competent professional should be sought.

Library of Congress Cataloging-in-Publication Data

Improving healthcare through built environment infrastructure / edited by Mike Kagioglou, Patricia Tzortzopoulos.
    p.   cm.
  Includes bibliographical references and index.
  ISBN 978-1-4051-5865-7 (hardback : alk. paper) 1. Health facilities—Design and construc-tion. 2. Health facilities—Great Britain—Design and construction.   3. Construction industry—Management—Research. 4. Infrastructure (Economics) I. Kagioglou, Mike.
II. Tzortzopoulos, Patricia.
  RA967.I47   2010
  725'.51—dc22
                                                                                          2009035244
A catalogue record for this book is available from the British Library.

Set in Sabon 10/12.5 by MPS Limited, A Macmillan Company
Printed and bound in Malaysia by Vivar Printing Sdn Bhd

1 2010

Contents

About the Editors

Mike Kagioglou is a Professor of Process Management and Head of the School of the Built Environment, University of Salford. He comes from an enginnering manufacturing background and for the past 12 years he has been undertaking research and teaching in the area of the built and human environment. He has worked on, and led, a number of research projects and initiatives including the 'Process Protocol' and managing the Salford IMRC (SCRI – Salford Centre for Research and Innovation) for its first five-year cycle, leading to its renewal until 2011. Mike is currently the Director of SCRI, the Academic Director for Salford University of the collaborative Health and Care Infrastructure Research and Innovation Centre (HaCIRIC – a collaborative IMRC between Salford, Loughborough, Reading and Imperial College), with particular interests in benefits realisation for healthcare, operations and lean management, and knowledge management through visible performance management. Mike has published over 100 academic and industrial papers and reports. Mike's overall drive is towards sustainable and enhanced healthcare outcomes for society through sustainable and innovative organisational and infrastructure-related structures and enablers.

   Patricia Tzortzopoulos is a senior lecturer having previously being an Academic Fellow at the School of the Built Environment, University of Salford, UK. She comes from an architectural background and has been developing research for over 10 years focusing on different aspects of design and the built environment. Her research interests cover design management; new product development; process management and operations management. She has worked on a number of research projects examining the design process and, for the past 5 years, focusing on healthcare facilities, looking at the design front-end issues, the effects of design on health outcomes and the links between buildings design and healthcare operations management. Patricia is currently a research coordinator in Salford for the collaborative Health and Care Infrastructure Research and Innovation Centre (HaCIRIC) and is associated with the Salford Centre for Research and Innovation (SCRI) in the Built and Human Environment. Patricia is also part of the advisory editorial board of the Journal of Design and Health.

About the Contributors

Bernard Aritua is a Research Associate and Doctoral Research Scholar at the School of Civil Engineering, University of Leeds. His PhD research involves investigating the management of risk and uncertainty in construction multi-project environments. The main objective of the research is to explore how the concepts of complexity science and systems dynamics may be used to enhance risk-based decision-making in programme and portfolio management. The principal focus of application is the procurement of major capital infrastructure in the public sector. Bernard's industry experience includes design and construction supervision of civil and building projects in Europe, Middle East and Africa. His project management experience has included both technical and management aspects of project appraisal and business strategy in civil engineering.

James Barlow holds a Chair in Technology and Innovation Management at the Imperial College Business School. Since June 2006 he has also acted as Principal Investigator and co-director of the Health and Care Infrastructure Research and Innovation Centre. James was educated at the London School of Economics. He has previously held posts at the University of Sussex, the Policy Studies Institute and the University of Westminster. His research focuses on innovation in complex sectors of the economy, with particular emphasis on healthcare and construction. He currently leads a major programme of research on the adoption, implementation and sustainability of innovation in healthcare systems. James also has extensive experience in providing policy and strategic advice to government and industry. He has published widely and is the author of over 100 papers or reports and four books.

Steffen Bayer is a Research Fellow in the Health Management Group at Imperial College Business School. He holds Master Degrees in Science and Technology Policy Studies from the University of Sussex and in Physics from the University of Texas. His doctoral research at the University of Sussex investigated environmental management in the steel industry. Prior to working at Imperial, Steffen was a Research Fellow at SPRU – Science and Technology Research, University of Sussex. His main research interests are innovation in healthcare, in particular home-based technology-supported health delivery (telecare), healthcare planning (e.g. stroke care) and the management of project-based organisations (e.g. architectural practices or engineering consultancies). He uses a variety of approaches in his research including system dynamics and discrete event simulation. Steffen is Imperial's research coordinator for the Health and Care Infrastructure Research and Innovation Centre and member of the policy council for the UK Chapter of the Systems Dynamics Society.

Denise Bower is a Professor of Engineering Project Management and the Deputy Head of the School of Civil Engineering at the University of Leeds. She is also a consultant to Oriel Group Practice, an internationally recognised project management consultancy. Her recent work includes the evaluation of procurement strategies, assessment of corporate strategy, the development of organisational partnering guidelines, the evaluation of the success criteria for a number of partnering arrangements and recommendations of contract strategies for overseas projects. Denise has a wide and varied experience of lecturing including international conferences such as IPMA, ECI and PMI and professional development courses for industries including manufacturing and oil and gas. She is also the Director of the MSc Engineering Project Management that attracts international students from a wide range of backgrounds. Denise is directing doctoral studies in the area of procurement and contracts, her particular area of research interest being in the optimisation of the procurement of contracted services. In recognition of her work she was a member of post-Latham Working Group 12, which produced the guide *Partnering in the Team*. Denise is the author and joint author of many books and publications, including *The Management of Procurement, Engineering Project Management, Dispute Resolution for Infra-Structure Projects and Managing Risk in Construction Projects.*

David Chambers is a registered architect and the Director of Planning and Architecture for the Sutter Health System, and has devoted the majority of his professional career to the healthcare field. For over 20 years he has evolved architectural planning concepts through envisioning optimised patient flows as the value stream, driving out waste and enhancing quality outcomes. He has developed this approach by bringing together high-level multidisciplinary care teams to map current state and then optimised patient flows, then applying these optimised flows to spatial programmes. This planning approach has been recognised for its high-value outcomes both nationally and internationally. In addition to overseeing several billion dollars of planning implementations for acute and ambulatory care centres, in the United States and Canada, Chambers is named as a source for the Healthcare Advisory Board and the Rice Building Institute. He consults with international healthcare authorities, presents at national symposiums and international congresses, and was named one of the "Twenty who are making a difference" in Healthcare Design Magazine as well as "Who's Who in Facility Management" for Facilities with more than 500 beds in Facility Care Magazine. His planning concepts have led to "one-stop" patient intake centres, integrated interventional services platforms and deeper implementation of acuity-adaptable inpatient nursing platforms.

Ricardo Codinhoto is a qualified architect with industrial, teaching and research experience. Ricardo holds a lectureship at the School of the Built Environment, having previously being a research fellow within the Health and Care Infrastructure Research and Innovation Centre (HaCIRIC) and Salford Centre for Research & Innovation (SCRI) at the University of Salford. He is active in a number of research initiatives related to design theory, management and practice in construction.

Richard Curry is an independent consultant specialising in telemedicine and telecare. He has authored several reports on aspects of telemedicine and telecare for the Department of Health, and was a member of the Telecare Policy Collaborative. He has also worked with a number of NHS Trusts, District Councils and Charities helping them to implement telemedicine and telecare programmes. He is currently Scientific Advisor to the Department of Health's research initiative on the role of technology in supporting chronic disease management, self care and healthy living. He is also project manager for a major EPSRC-funded project on the planning and implementation of telecare. He has published extensively on the subject of telecare with James Barlow and Steffen Bayer. Prior to working as a consultant he was Business Development Manager of a videoconferencing company where he designed and installed the Minimally Invasive Therapy Training System based at the Royal College of Surgeons. He holds a PhD in physics from the University of Durham.

Andrew Dainty worked briefly as a site engineer and as a design engineer, and then studied for his PhD at Loughborough University in which he investigated the career dynamics of men and women working within large construction organisations. On completion of his PhD, he joined Coventry University as a Lecturer and then Senior Lecturer in Construction Management. He rejoined Loughborough in 2001 and is now Professor of Construction Sociology. His research interests focus on human social action within construction and other project-based sectors, and particularly the social rules and processes that affect people working in construction teams, organisations and supply chains. He holds several research grants in which he is exploring various aspects of the management of people in the industry.

Ged Devereux started work for Manchester City Council in 1998 as part of a specialist health promotion unit set up to strengthen the Council's public health role. He was appointed to his current post in April 2003, and now manages a programme of work that incorporates partnership working to develop and maximise the built environment for healthcare. The Unit has been established to lead and support work on reducing health inequalities in Manchester, and between Manchester and the rest of the country. The Unit is funded by NHS Manchester and Manchester City Council and is based in Manchester Town Hall. Ged has worked in Manchester for 20 years and has held jobs with Educational Services, Environmental Health, Regeneration and Social Services before moving across to the Joint Health Unit in 2003. During this time he has managed a range of services and built environment-related programmes including those with a focus on regeneration such as the Manchester, Salford and Trafford Local Investment Finance Trust (LIFT) local hospital Private Finance Investment (PFI) schemes and has contributed to the development of Strategic Regeneration Frameworks to regenerate deprived communities in Manchester. Ged completed his post-graduate degree in Health Care Management in 2001 at Manchester University; in June 2008 he completed a Masters of Science in Built Environment for Health Care at Salford University.

Susan Francis is Special Advisor for Health at CABE and works closely with the UK Government's Department of Health for whom she chairs the NHS Design Review Panels for all major investment projects. CABE is the Commission for Architecture and the Built Environment, the UK government's advisor on architecture, urban design and public space. Susan works with CABE to support health clients to develop excellent design briefs, generate and publish information on the benefits of good design, and pioneer new thinking about healthcare design. Susan is an invited member of the Sustainable Development Commission Expert Panel and chairs the prestigious Building Better Healthcare Awards. Trained as an architect, Susan has worked in this specialised field for nearly 20 years: as an academic at MARU (Medical Architecture Research Unit) developing research and post-graduate training; as design lead for the Future Healthcare Network for over 70 NHS Trusts developing capital programmes; with the King's Fund programme to enhance existing buildings; and with the UK government developing policy on supporting and evaluating design quality. Susan has published and presented extensively in the UK and abroad.

Richard L. Groome is a Chartered Chemical Engineer and a Fellow of the Royal Society of Public Health; currently Operations Director for John Laing plc in the Central area of the UK. In a career spanning 35 years, his early roles were in the food and dairy industries, where he was responsible for many of the innovations in the latter that are now commonplace; such as preservative- and additive-free foods, portion packs, pots with separate compartments etc. Later he became involved with major construction projects, particularly where high-level specifications and finishes were required, and also served as a non-executive Director of a Health Authority for 6 years. He became very interested in fast track building solutions and off-site construction, and was heavily involved in the rollout of the UK mobile telephone networks during the 1990s using such methods. By 2002, he was developing these techniques to provide fast track health centres, clinics and surgeries. In 2005 he became Chief Executive of the Manchester Salford and Trafford LIFT Company, a Private Public Partnership to deliver healthcare infrastructure solutions for 25 years to Greater Manchester, the largest such operation in the UK, and delivered £80 million worth of new projects within 3 years. For the past year he has had operational responsibility for 11 larger projects delivered under the Government's Private Finance Initiative (PFI), which cover such diverse requirements as police stations, hospitals, street lighting and schools; and he holds 20 Directorships on Special Purpose Companies with a wide variety of clients.

Jane Hendy is a part of the Healthcare Management Group and member of the Health and Care Infrastructure Research and Innovation Centre (HaCIRIC). Jane previously held positions at University College London, and the London School of Hygiene and Tropical Medicine. She has an MSc and PhD from the University of Surrey in Applied Psychology. Jane's research focuses on organisational change and the adoption of innovations in healthcare. She is particularly interested in large, potentially transformational, organisational initiatives. Previous work included conducting the first in-depth exploration of the £12bn NHS National Programme for Information Technology. Current research includes a £150m government initiative to introduce technology-based health and social care services (telecare) to help support people with chronic conditions. Other work includes a £30m Department of Health initiative to accelerate the application of health research into better care. Jane's work focuses on organisational factors underlying the successful and sustainable introduction of these new methods of health service delivery.

Ahmed Doko Ibrahim obtained PhD at Loughborough University, studying continuous improvement in the procurement of healthcare facilities. He has worked previously as Project Quantity Surveyor in consulting firms, as a Research Assistant at King Fahd University of Petroleum and Minerals and as a Research Associate in the Health and Care Infrastructure Research Innovation Centre (HaCIRIC) at Loughborough University. Ahmed is currently a Senior Lecturer and Head of Department of Quantity Surveying, Ahmadu Bello University, Zaria, Nigeria. His current teaching and research interests are in cost modelling, public–private partnerships, innovative procurement of healthcare facilities and continuous improvement.

Sarel Lavy is Assistant Professor at the Department of Construction Science, College of Architecture, Texas A&M University. Dr. Lavy graduated with a PhD in Civil and Environmental Engineering from the Technion – The Israel Institute of Technology, and has been a faculty member in the Department of Construction Science since 2005. The Department of Construction Science is one of four departments in the College of Architecture at Texas A&M University. Dr. Lavy also serves as the Associate Director and a Fellow of the CRS Center for Leadership and Management in the Design and Construction Industry, as well as a Fellow of the Center for Health Systems and Design. Dr. Lavy is a member of the International Facility Management Association (IFMA), the International Council for Research and Innovation in Building and Construction (CIB), Working Commission W070 – Facilities Maintenance and Management, the American Society of Civil Engineers (ASCE), and the American Society for Healthcare Engineering (ASHE). Dr. Lavy's principal research interests are facilities management in the healthcare and education sectors, maintenance management, and performance and condition assessment of buildings. His previous research studies led him to author and co-author a total of 11 papers published in peer-reviewed journals and 15 papers published in peer-reviewed conference proceedings in international, national and local venues. He teaches facility management graduate classes at Texas A&M University, and serves as a member on the Facility Management Certificate Council, which is the authoritative body that administers and governs the facility management certificates.

Therese Lawlor-Wright is a Lecturer in Project Management in the School of Mechanical, Aerospace and Civil Engineering at the University of Manchester. She holds a degree in Industrial Engineering from the National University of Ireland Galway, an MSc in Advanced Manufacturing and a PhD in Mechanical Engineering from the University of Manchester. She has worked in industry and academia as a project manager and knowledge management consultant for over 20 years, completing successful projects with multinational and local engineering companies. Her research area includes knowledge management in design and she has led several research and knowledge transfer projects in the area of Concurrent Engineering. With experience in diverse sectors, Therese's research interests are in the management of complex projects – particularly in performance improvement in multidisciplinary and multicultural environments.

William A. Lichtig is a shareholder in the Sacramento-based law firm of McDonough Holland & Allen PC, USA, where his practice focuses on construction, surety and public contract law. A construction lawyer for more than 20 years, Will has seen first hand some of the fundamental breakdowns in traditional project delivery. As special counsel to Sutter Health, a major Northern California healthcare organisation, he provides a full range of legal services in support of its $6.5+bn design and construction programme. In a similar capacity, Will serves as special construction counsel for the California Prison Health Care Receivership's $8bn design and construction programme to update and expand the state prison medical infrastructure. Will has been instrumental in instituting innovative project delivery solutions for Sutter Health and others, including Lean Project Delivery. He has developed unique supporting documents such as the Integrated Agreement for Lean Project Delivery and an incentive fee plan. His innovations and contributions to the industry were recently acknowledged by *Engineering News-Record* magazine when he was named an 'ENR Top 25 Newsmaker' for 2007. In 2009, Will was selected by his peers for inclusion in *The Best Lawyers in America®* Construction Law list.

Jose Barreiro Lima comes from an engineering and management background. Over the past years he delivered consulting services in Portugal and abroad, with a special focus on the construction industry. He also has teaching experience, with a special interest in the exciting and dynamic intersection between technology and business management. José holds a Civil Engineering degree from the Portuguese University of Minho and an MBA from the Oporto Business School. He also holds an MSc Management from University of Oporto, where he developed a dissertation on 'Performance and Strategies – Analysis of Leading Contractors in Portugal'. In 2008, he obtained a PhD degree from the Salford University's Research Institute for the Built and Human Environment in 'Methodology for

Demand-Supply Selection of Commercial Off-The-Shelf Software-Based Systems – Contextual Approach of Leading Contractors in Portugal'. Since 2008, José is a Researcher Fellow at the Health and Care Infrastructure Research and Innovation Centre (HaCIRIC) at the University of Salford. In Portugal, José lectures Construction Project Management in the University of Minho Master' courses and is the Regent's Professor (Invited) of Information Technology for the Marketing and Management courses, at the ISG – Management School.

Laurie McMahon is a Professor in Health Policy at City University, London and Director of Loop2. He has extensive experience in management and organisational development and change across a broad range of public and private sector organisations both in the UK and overseas. He is the co-founder of The Office for Public Management and, before forming Loop2, led their healthcare practice. He was a Fellow of the King's Fund and a senior consultant in overseas healthcare. His main interests are in strategy development and implementation, organisational design and delivery, engineering large-scale organisational change and the use of behavioural modelling to understand complex 'futures'. Recently he has focused on helping provider and commissioner organisations respond to the introduction of market forces into the NHS. He is also Policy Advisor to Nuffield Hospitals, Special Adviser for the WHO Office for 'Investment for Health', a Fellow of the Institute of Quality Management and a Fellow of the Health Finance Management Association.

Duane Passman, originally an astrophysicist, has worked on major investment programmes and projects in the NHS for over 20 years. These projects included the Chelsea & Westminster Hospital, various developments at St. Thomas' Hospital in London, the redevelopment of Chapel Allerton Hospital in Leeds, the Jubilee Wing at Leeds General Infirmary, the award-winning £238m Queen's Hospital in Romford and the redevelopment of the North Middlesex Hospital. He was a contributor to the 1994 Capital Investment Manual, and has worked on the development of a significant number of large investment business cases, including developments in Leeds, Bradford, Manchester, Birmingham and the island of Malta. Duane was also Head of Capital Investment for the NHS in London from 2001 to 2007 and provided advice, guidance and support to all elements of the NHS capital investment portfolio across the capital. He joined Brighton & Sussex University Hospitals NHS Trust in August 2008 as the Programme Director for the development of teaching, trauma and tertiary services, which will result in a £300m + redevelopment. He is a Visiting Professor in the School of the Built Environment at the University of Salford, and is a member of the Salford Centre for Research and Innovation Steering Committee. Duane has published and presented widely on healthcare planning and capital investment in the UK and internationally.

Bronwyn Platten is an Australian artist who has exhibited her artwork internationally in numerous group and solo exhibitions. Most recently her work has been included in an international touring exhibition 'Figuring Landscapes', which was launched at the Tate Modern, London in 2009. She has worked in a number of universities as a lecturer and researcher including Gray's School of Art, The Robert Gordon University, Aberdeen: South Australian School of Art, University of South Australia and Canberra School of Art and The Australia National University. Platten has received numerous awards for her art practice from the Australia Council for the Arts, and in 2004, she received an AHRC grant to develop the international research project 'Imaginal Regions' www.imaginalregions.co.uk. Since 1985 she has also worked with diverse community groups to establish creative art projects, most notably the award-winning 'Building Art Project' (1997) in which, with artist John Foubister, she worked collaboratively with 25 artists with an intellectual disability. From 2001 she has been involved intensively in arts in hospitals working as both an artist and a curator for Grampian Hospitals Art Trust, Aberdeen Royal Infirmary and subsequently as an Arts Project Manager for Lime, an arts in health organisation that develops creative arts practice in hospitals across Greater Manchester. She is currently undertaking a PhD within HaCIRIC at the University of Salford researching the relationship between patient well-being and creativity utilising a participatory arts-based approach.

Andrew Price has over 25 years of experience in design, construction and industry-focused research. He obtained BSc in Civil Engineering from Nottingham Trent University. He has worked for four years as Structural Engineer for Jackson Peplow Consultants before joining Loughborough University as Research Assistant in 1981. He became a lecturer in Construction Management in 1984, and is currently a Professor of Project Management in the Department of Civil and Building Engineering. His early research focussed on construction productivity and the motivation and development of human resources. This evolved to include several project management-related topics, including integrated design and construction, integrated supply chains, partnering and less adversarial long-term relationships. In recent years, the focus has moved towards measuring and improving the socio-economic aspects of construction performance, which included: construction value, sustainability; performance improvement; Total Quality Management; and benchmarking. Current research includes innovative design and construction solutions for health and care infrastructure, continuous improvement and sustainable urban environments.

Stelios Sapountzis is a research fellow at the Health and Care Infrastructure Research and Innovation Centre (HaCIRIC) at University of Salford. He comes from a manufacturing background having attained a BSc (Hons) in Manufacturing Management and has postgraduate studies in Advanced Manufacturing Systems. He has spent 10 years with a world-class contract electronics manufacturer as a production and operations manager focusing on process improvements and lean manufacturing implementation. Stelios's research interests are in the field of Process and Change management and Lean service delivery. He is currently working towards a PhD developing a benefits realisation framework for healthcare infrastructures and services.

Igal M. Shohet is Associate Professor, Head of the Construction Management Program, Deputy Chair of the Department of Structural Engineering, Faculty of Engineering Sciences, Ben-Gurion University of the Negev and Chairman of the Construction Management Division of the Israel Society of Civil Engineers (ISCE). Dr. Shohet earned his degrees from the Technion – Israel Institute of Technology: BSc in Civil Engineering (Cum Laude); MSc in Construction Management; and DSc in Civil Engineering. Dr. Shohet was appointed visiting professor at the University of New Mexico (1995–1996), and at the Georgia Institute of Technology (2004), and was a faculty member at the Technion's Department of Civil and Environmental Engineering (1996–2004) and National Building Research Institute (1994–2004). In the past two decades, Dr. Shohet performed research in the areas of construction management, construction robotics, construction methods and facility management. Dr. Shohet has published over 70 refereed articles in peer-reviewed professional journals and in conference proceedings. Dr. Shohet's key performance indicators for healthcare facilities, developed over the past decade with his colleague Dr. Sarel Lavy, are widely implemented in Israel in the public healthcare sector, and in the public sector in Israel and in the USA. Dr. Shohet is a member of the CIB W70 Working Commission for Facilities Management and Maintenance and is a referee for leading journals in the area of construction and facilities management, namely Construction Management and Economics, International Journal of Strategic Property Management and Journal of Building Appraisal. In 2004, Dr. Shohet joined the Structural Engineering Department at BGU and established the Construction Management programme. The new programme, together with the department itself that was established in 2002, has an undergraduate student body of 300 in the areas of structural engineering and construction management. The programme offers unique concepts in engineering education, focusing on extreme events mitigation in the built environment and a comprehensive internship programme in structural engineering and construction management.

Nigel Smith is Professor of Project and Transport Infrastructure Management and the Head of the School of Civil Engineering at the University of Leeds. After graduating from the University of Birmingham, he spent 17 years in industry principally working on major transport infrastructure projects in the UK. Since returning to university his research interests have included procurement methods involving private finance, risk management and resilient infrastructure. Nigel maintains close links with the Institution of Civil Engineers and is on several committees including the Editorial Board of the Management, Procurement and Law Journal. He is also on the Editorial Boards of the International Journal of Project Management and the Journal of Engineering Construction and Architectural Management. He has published widely in international refereed journals and is the author/editor of several books including, Engineering Project Management, Managing Risk in Construction Projects and Appraisal Risk and Uncertainty.

Kate Trant is a design historian, and Senior Research Advisor for the Commission for Architecture and the Built Environment, the government's advisor on architecture, urban design and public space, working closely on CABE's health and well-being research and strategy. She worked with the Design Brief Working Group on 'Advice: the main components of the design brief for healthcare buildings', and 'Dying with Dignity' for the Department of Health, and is a member of the NHS Design Review Panel. She convened CABE's first Health Week in 2006, which brought together a wide range of individuals and organisations across health and well-being to take a detailed look at the relationship between health and the built environment. Beyond the arena of health and well-being, Kate is a writer and exhibition curator, and has worked with organisations including the British Museum, the Royal College of Art, the Science Museum, the New Art Gallery Walsall and the Royal Institute of British Architects. She curated *Moving Objects: Thirty Years of Vehicle Design* at the Royal College of Art in London, and the Interpretation Centre for the New Art Gallery Walsall, during its construction. In 2008, she commissioned a major piece of public art for CABE's first Climate Change Festival, which took place in Birmingham – a full-scale, mirrored pylon, installed in the city centre to bring home the reality of climate change in our towns and cities. She has written articles for a wide range of architecture, design and transport publications, and has co-authored two books: *The Macro World of Microcars*, and *Home away from Home: The World of Camper Vans and Motorhomes*.

Kathryn Yates is a researcher at the Health and Care Infrastructure Research and Innovation Centre (HaCIRIC) at the University of Salford. Her background is in Sociology in which she has a BSc (Hons) from Sheffield University. She completed a Masters in Healthcare and the Built Environment at the University of Salford in 2008, with a dissertation focusing on the evaluation of a cross-disciplinary and multi-stakeholder teams. Kathryn is conducting research on Health and Care infrastructures and project/programme management issues related to the built environment, with particular focus on the areas of benefits realisation, stakeholder management, communication, team work and leadership. She has a strong theoretical understanding of these areas and has used this to aid the development of a Benefits Realisation Framework and Process (BeReal); and facilitate Benefits Realisation workshops for Local Improvement Finance Trust (LIFT) schemes, Primary Care Trusts (PCTs) and the Acute Sector.

Foreword

The impact of the built environment on people has been studied, written about and taught for as long as written records have existed. Indeed the impact that some of the world's oldest built environments still have today demonstrates that the people that created them understood the impact that they would have.

In recent years, surveys by the Department of Health in England, and research and studies around the globe have confirmed long-held beliefs that the built environment is directly linked to a range of factors that impact on the successful delivery of healthcare. These include confidence in the organisation using the environment to deliver services, feelings of security, privacy, dignity and well-being, efficient working, the ability to respond to changing requirements and the ability for people to communicate effectively in stressful circumstances. The surveys and studies have also shown that the design of the built environments of previous times do not efficiently or effectively meet the desires or needs for the healthcare environments of today or tomorrow.

I have been fortunate to work on the planning, design, delivery, operation and management of built environments to support a wide range of activities, including healthcare delivery, for over 30 years. I have experienced numerous examples where well-researched, planned and executed new or changed environments have improved the organisation's ability to deliver services and the experience of the people receiving the services, visiting or living close to the environment. Sadly, I have also been complicit in delivering built environments that did not provide these outcomes. Whilst I carry those scars with me and strive never to repeat the mistakes that led to them, I have no means of effectively passing on the lessons.

The built environment for the delivery of healthcare will continue to change as it responds to new technologies and modalities of care, different expectations and requirements of providers and consumers of care. It is vital that built environment students and practitioners alike avail themselves of the best possible information to guide them in their studies, continuing professional development and the delivery of their tasks. The range is enormous from the assessment of need, planning the service delivery to design, construction, commissioning, maintenance and operation of the healthcare environment.

The book that follows addresses these areas from a blend of contributions of experienced practitioners to the descriptions of the output from recent research that moves forward the frontiers of knowledge and practice in the many areas of the healthcare built environment.

I happily commend this book to all engaged in the exciting fields of planning, delivering, maintaining and operating healthcare environments. When we get it right, we are able to do immeasurable good.

Rob Smith
Director of Estates and Facilities (NHS England), Department of Health

Acknowledgements

This book would not have existed without the hard work of all the authors, editors and the reviewers who peer-reviewed all chapters. Special thanks go to all organisations who are striving for better health and care delivery and to those practitioners who are courageous in approaching academia and experimenting on improved ways and new paradigms through the co-production of knowledge.

Introduction

Improving Healthcare through Built Environment Infrastructure

Mike Kagioglou and Patricia Tzortzopoulos

Healthcare service provision is increasingly becoming a complex and dynamic process affected by myriad causes that cut across socio-techno-economic boundaries. The fast pace of policy and structural changes across private and public organisations, combined with an accelerated change in technological innovations, has challenged traditional viewpoints and paradigms in how health and care services are conceived, designed, implemented and sustained in what has become a very expensive and resource-intensive environment. Innovation is no longer an optional extra but part of the essence of any changes suggested for now and the future – innovation needs to be normal practice.

There is a growing drive and determination from public sector bodies worldwide to improve the populations' health, including their physical and mental well-being. There is also a clear recognition that the built environment in general and healthcare infrastructure in particular have a very important role to play in terms of supporting health and well-being, encouraging healthy lifestyles, improving quality of life and increasing efficiency and effectiveness. In the UK, healthcare challenges arise from issues like an ageing population that needs better, more personalised healthcare as well as more supportive environments to be enabled to live longer independently. This imposes new challenges for the built infrastructure. The impact of the physical environment on health and well-being is highlighted in many reports including a foresight programme that will advise the UK government on how to achieve the best possible mental development and well-being for everyone in the future (see http://www. foresight.gov.uk). Consequently, new healthcare initiatives focus on community settings, seeking conditions to provide integrated, personalised healthcare services closer to the communities and home, widely and sustainably supporting prevention and healthy living styles, and supporting the regeneration of urban areas. Indeed, the increasing awareness of the role of the built environment for health and well-being resulted in it being at the heart of many public sector initiatives across the UK.

Despite the growing interest in healthcare infrastructure, it remains difficult for both practitioners and academics to be informed about current practices and advances. The literature on healthcare infrastructure is spread across numerous journals, conference proceedings, magazines and focused books in a variety of disciplinary areas. Furthermore, a range of perspectives exist, looking at the role of the environment in health from a broad range of areas like design, psychology, psychological studies, social sciences, health and social care studies, to mention a few. As such, there is no single understanding or a dominant paradigm on how to improve healthcare through the built environment infrastructure.

Past books in the area of healthcare infrastructure have concentrated on the medical or design issues of buildings but none has considered adequately the 'meeting space' of management science and built environment technologies with the medical and operational needs of healthcare settings, including future-proofing and considerations of healthcare models of delivery. The purpose of this book is to address some of the key issues related to healthcare delivery with regard to capital investment and infrastructure, design and operations management theories and their application to healthcare. These are looked at in a variety of settings such as primary, acute and home-based. The book is more specifically positioned to address the issues related to the devolution of health and care from acute and secondary care to the primary, intermediary and home care. It also aims to address issues related to public–private partnerships and how these operate in the context of healthcare delivery.

The book brings into one place a compilation of the diverse research and practical examples of improving healthcare delivery through the built environment. We hope this book serves as a useful reference for practitioners and academics who seek to understand and advance the area of built environment for health. This book takes a multidisciplinary approach, with contributions from both theory and practice. Consequently, it has been divided into two main sections: practitioner and academic contributions.

We are aware that there have been aspects of healthcare and the built environment that are not covered in this book. However, we believe that the main thrust of research and practice has been approached. Finally, the book has benefited from international contributions, aiming to keep a global perspective on the challenges around the planning, design, construction and maintenance of healthcare infrastructure.

1.1 Part 1: Practitioner contributions

This part brings practitioner's perspectives and examples of issues and solutions related to the planning, design and delivery of healthcare facilities. Most contributions are from the UK, with the addition of perspectives of improving healthcare through the built environment from the US context.

Chapter 2, by Duane Passman, provides a broad view of planning healthcare environments in the UK. It presents a historical review of investment in the National

Health Service (NHS) built infrastructure, discussing political healthcare agendas and how these have been influencing the design and delivery of facilities. The chapter also describes methods available to the NHS to procure healthcare facilities, emphasising the needs of different organisations across the NHS. The chapter concludes by briefly approaching the importance of good design for healthcare.

Chapter 3, by Sue Francis, centres around designing healthcare environments for change and flexibility. The chapter discusses healthcare policies, changes in policies and how these influence the built environment. The different places where health is delivered, from home to highly specialised hospitals, are presented. Finally, the chapter concentrates on how essential design is for the improvement of services and for staff and patient satisfaction. It also describes design quality measures. The chapter presents examples of award-winning hospital designs.

Chapter 4, by Kate Trant, looks at the role of design in creating individual buildings, as well as how the design of high-quality places contributes to creating healthy neighbourhoods. It discusses some tangible and intangible characteristics of the environment and how they might influence well-being and quality of care. This chapter also presents examples of award-winning hospital designs, which contribute to the creation of healthy neighbourhoods.

Chapter 5, by Richard Groome, describes the process stages in the delivery of primary healthcare facilities through the UK's LIFT (Local Finance Improvement Trust) procurement. The chapter discusses issues around the different design stages, going through to construction and facilities management. It finalises by bringing to light some cultural differences between public and private sector bodies, and tensions that such differences impose in long-term public–private partnerships such as LIFT.

Chapter 6, by William Lichtig, brings a lean perspective to the delivery of healthcare facilities from a North American perspective. The chapter presents concepts related to lean project delivery and an integrated form of agreement, which are being adopted by a healthcare organisation in California, Sutter Health. It discusses issues around the creation of a collaborative environment for design and construction through a network of commitments, aiming at achieving maximum value for patients and staff.

The final practitioners' contribution in Chapter 7, by Dave Chambers, continues describing experiences from North America through innovations led by Sutter Health. The chapter centres on a prototype hospital initiative, discussing issues about improving service delivery and patient flows as well as delivering new facilities, and briefly discusses issues around setting goals and metrics and achieving these through building design.

1.2 Part 2: Academic contributions

Part two presents results from current research looking at diverse aspects of the healthcare built environment. Knowledge from a broad range of areas is discussed, from strategy development and the devolution of care from acute to primary settings and telecare, through to risk management in procurement, design, benefits realisation, continuous improvement, performance management and facilities management in healthcare environments.

Chapter 8, by Ged Devereux, examines the effectiveness of the Strategic Service Development Plan (SSDP) to support the development of built environment solutions for primary healthcare services in the UK. The chapter analyses the development of SSDPs at three UK localities, which are public sector partners forming the Manchester, Salford and Trafford MaST LIFT. The analysis focuses around partnership working, the planning process itself and the realisation of benefits from the projects delivered to the communities.

Chapter 9, by James Barlow, Steffen Bayer, Richard Curry, Jane Hendy and Laurie McMahon, discusses issues around the devolution of care from acute hospitals to primary care and ultimately home, focusing around telecare and the impact it may have on future infrastructure requirements.

Chapter 10, by Nigel Smith, Denise Bower and Bernard Aritua, presents an overview of the management of risk throughout the NHS procurement process. The chapter explores collaborative procurement, as well as multi-project approaches for the delivery of healthcare facilities. It concludes by considering views for the future for sustainable procurement practice.

Chapter 11, by Ricardo Codinhoto, Bronwyn Platten, Patricia Tzortzopoulos and Mike Kagioglou, describes challenges and issues around the creation and adoption of an evidence-base to inform healthcare design development and evaluation. The chapter presents a framework bringing together knowledge from diverse and dispersed research into a single place, to facilitate its adoption in practice.

Chapter 12, by Stelios Sapountzis, Kathryn Yates, Jose Barreiro Lima and Mike Kagioglou, describes a benefits realisation framework developed to support the management of programmes and projects for the delivery of healthcare facilities.

Chapter 13, by Ahmed Ibrahim, Andrew Price and Andrew Dainty, proposes a framework for continuous improvement for the UK Local Improvement Finance Trust (LIFT) initiative. The chapter centres around concepts of continuous improvement and how these can be applied to support LIFT through time.

Chapter 14, by Therese Lawlor-Wright and Mike Kagioglou, focuses on performance management issues and how these have been approached within NHS organisations. The chapter examines issues around performance of healthcare facilities from design through to operation, and concludes with a discussion of how infrastructure may contribute to the performance of healthcare organisations.

Chapter 15, by Igal Sohet and Sarel Lavy, synthesises the state of the art in hard healthcare facilities management, presenting its different components, as well as key performance indicators. The chapter then examines a case study illustrating the use of the proposed indicators in a hospital setting in Israel. The chapter concludes by offering a view towards a performance maintenance toolkit.

The final chapter, 16, by Igal Sohet, considers the core facilities management concepts of community clinic facilities and compares these with hospital facilities. The chapter presents key performance indicators for community clinic facilities and tests these through a case study.

Part 1

Practitioner contributions

Planning Healthcare Environments

Duane Passman

2

2.1 Introduction

The purpose of this chapter is to identify some of the key drivers that have an impact on the planning and delivery of healthcare environments. It mainly examines the English NHS system, but there are more similarities across the four UK healthcare systems than there are differences.

This chapter will not attempt to rehearse or provide a detailed history of the planning of healthcare environments, as there are other sources available that deal with this subject in more depth than can be attempted here. It does attempt to give a flavour of what needs to be considered and how these can be applied in what can be a challenging planning environment subject to rapid change.

2.2 Background and history

The website of the NHS History (see http://www.nhshistory.net) provides an interesting history of hospital planning and development, which not only focuses on the London healthcare system primarily, but also identifies the interaction between national policy and the London system. In many ways, the development of the London healthcare system has strongly reflected the development of the national system, and many of the debates about primary, secondary and tertiary care and between-hospital and out-of-hospital care are most strongly represented in the debates about NHS provision in the capital.

As the NHS recently celebrated its 60th anniversary, the Department of Health (DoH) also summarised some of the key dates and milestones achieved in NHS history since 1948 (see http://www.nhs.uk/aboutnhs/nhshistory/Pages/TheNHShistory1960s.aspx). The creation of the NHS in 1948 brought about the transfer of a diverse collection of healthcare buildings (related to size, condition, age and functionality) into public ownership. Until around 1962, the

NHS was governed at the local level by a multitude of different boards and committees that were rarely collaborative or cooperative – even between neighbouring authorities.

2.2.1 The hospital plan of the 1960s

In 1962, the then Minister for Health, Enoch Powell, called for the separation of the NHS into three main divisions – local health authorities, general practice and hospitals. The Hospital Plan of 1962 was, as said by Powell, 'the opportunity to plan the hospital system on a scale which is not possible anywhere else certainly this side of the iron curtain'.

The plan proposed the creation of the district general hospital (DGH) – hospitals that would serve a population of 100 000–150 000 and typically would contain 600–800 beds. These would provide general, secondary care facilities with Accident and Emergency (A&E) departments to the local population. Complex, tertiary work would be undertaken by the larger teaching hospitals – which in most cases would also provide the DGH service to the areas in which they were located.

The plan was based on the aggregation of regional strategies for health service provision and investment. It is commonly felt that the figures that were produced for investment in the physical infrastructure – to create (in some cases) and to upgrade and extend existing facilities – were unrealistic, and by 1966 the plan was downgraded to a 'Hospital Building Programme'. The intervening 3 years provided a fraction of the investment that had been originally pledged, which is not surprising given the general nature of the economy and the construction industry at that time.

2.2.2 The economic crisis of the 1970s

By 1973, the oil crisis and the general global economic situation had almost completely halted any pretensions towards large-scale capital investment in the infrastructure of the NHS, which meant that much of the building stock that the NHS had inherited in 1948 was, increasingly, in poor and deteriorating condition and not fit for purpose. The prevailing economic climate and the unceasing ability of the NHS collectively to 'make do and mend' meant that although the nature and state of the estate was always a concern, it was not necessarily at the top of the operational or political agenda for the NHS. The lessons of history should be noted as the UK and world economies enter, in 2009, a potentially prolonged recession that may well match the impact of the 1970s on infrastructure renewal and replacement.

2.2.3 Change in the 1980s

During the late 1980s, the improvement in the economic climate meant that there was more public capital generally available for the NHS, but this increase was from a generally low base. Regional health authorities (RHAs) controlled the supply of major capital by and managed major developments. Although on one hand this meant that there were a collection of departments and individuals who were skilled in the development, design and execution of major investment programmes and projects, these were felt to be too remote from the hospitals that were undertaking the development – especially after the rise of managerialism at local level following the review of hospital and healthcare administration and management (DHSS, 1984), known as the Griffiths Report.

At that time, capital was allocated in the RHA 10-year plan, which was itself a combination of top-down and bottom-up planning from the districts and units within each RHA area. The management tier between the RHAs and the individual hospital units were the district health authorities, who had their own capital allocations, but these were relatively minor when set in the context of the major investment usually required for a typical DGH.

Each RHA prioritised where the available capital would be expended in a particular year and also took account of the level of surplus land and buildings that could be sold to provide additional capital resource, particularly following the publication of the Ceri Davies Report (DHSS, 1983) that found that the NHS had generally more land and buildings than it required and some could be released for development. The late 1980s and early 1990s saw an acceleration of the trend to release land and buildings with a concomitant trend towards rationalisation of smaller hospitals towards the DGH and teaching hospitals.

The zenith of this trend was the development of the Chelsea and Westminster Hospital in West London – developed by the removal of St. Stephen's Hospital on the Fulham Road and the construction of a 650-bed teaching hospital – the first purpose-built, entirely new teaching hospital in London for over 30 years and one of the largest hospital building projects in Western Europe. The development was planned to be funded from RHA capital, but this capital would be repaid to the RHA capital programme by the disposal of four hospital sites (Westminster, Westminster Children's, St. Mary Abbott's and Hammersmith Hospital for Women) at the end of the period.

However, the property slump of the early 1990s drastically reduced the potential sale proceeds, meaning that the majority of the cost of the development (£210m at 1992–1993 prices) had to be funded from the RHA capital programme. The impact of this was that the rest of the RHA capital programme in North West Thames was delayed for several years or projects were cancelled.

At the same time, the phase 3 redevelopment of Guy's Hospital was running into significant problems during the construction phase, with costs escalating dramatically and the project finally delivered almost 4 years late and £115m over the original budget of £35m. The NHS capital programme became a significant political issue with investigations undertaken by the National Audit Office (NAO) and the Committee of Public Accounts of the UK Parliament (known as the PAC) into these two projects. (For the Public Accounts Committee report on Guy's Phase 3, see http://www.publications.parliament.uk/pa/cm199899/cmselect/cmpubacc/289z/289z02.htm#evidence.)

As property prices slumped and the economy slowed, the number of major publicly funded projects once again slowed down dramatically. However, there was an increasing public and political perception that a good proportion of the NHS infrastructure – particularly in respect of the acute hospitals – was beyond its useful life.

The Conservative government had recognised that the demand for investment in public infrastructure – roads, water supply etc. – far outstripped the available supply of capital to address the historic under-funding in these infrastructure assets. These had not been addressed by successive governments as the general prevailing postwar economic conditions and costs of essential infrastructure reprovision meant that the backlog of maintenance in healthcare facilities was becoming a serious operational issue.

2.2.4 Further change in the 1990s

The late 1980s and early 1990s saw a major restructuring of the way in which the NHS was managed, following the publication of the White Paper 'Working for Patients' in 1989 (DoH, 1989). The implementation of this policy saw the creation of NHS Trusts as self-governing entities responsible for managing their own affairs – including the responsibility for managing major capital investment programmes and projects. At the same time, the RHAs divested of their design and project management functions. Many of the staff involved transferred into private practice, with others moving to those of the newly established NHS Trusts that were sponsoring major capital projects.

As part of the reforms introduced by 'Working for Patients', NHS organisations would have to explicitly acknowledge the cost of capital investment through 'capital charges', whereas capital had been considered a free good uptil that point. For the first time, the NHS would have to consider more carefully how capital was deployed and the rate of return expected from that capital. Although the detail of calculating capital charges has changed over that period, the principle remains in place in 2009 and is still an important factor in differentiating between priority investment areas for NHS organisations.

In 1994, the Capital Investment Manual (CIM) was launched by the NHS Executive (NHS Executive, 1994). The purpose of this was to address to critical NAO/PAC reports into the NHS major projects noted above and to address the shift of the management of these projects from the RHA tier down to individual NHS Trusts (see http://www.dh.gov.uk/en/Publicationsandstatistics/Publications/PublicationsPolicyAndGuidance/DH_4119896).

On a purely anecdotal basis, the CIM was seen across government as most closely aligned with the available best practice at the time, and many of the principles contained within it hold true today. The CIM focused mainly on delivery of schemes through public capital, which was still the key funding mechanism for NHS capital projects, although one of the sections dealt with the application of private finance – mainly leasing of medical and scientific equipment.

It was not apparent at that time what a key role private finance would have in the development of new NHS facilities in the 15 years that followed. The following sessions will provide an overview of health and health investment policy changes during the late 1990s up to early 2009.

2.3 The planning landscape

A key text that examines in detail the determinants of planning in the acute care environment and looks at the main drivers in supply and demand is Acute Futures (Harrison and Prentice, 1996). As many other publications by the King's fund, this book remains germane and relevant well over 10 years after its publication in 1996. It examines the history and trends in acute hospital planning and identifies key issues for the future – all of these issues remain valid today.

The overarching issue that drives how healthcare environments are planned is the application of policy. Policy is intertwined into all the elements of supply and demand in the NHS and, periodically, changes the priorities for the NHS. Figure 2.1 summarises some of the key determinants in planning healthcare provision.

The next section deals with some of the key developments and changes in policy over the past 10 years. Section 2.5 discusses the **capital procurement methodologies**

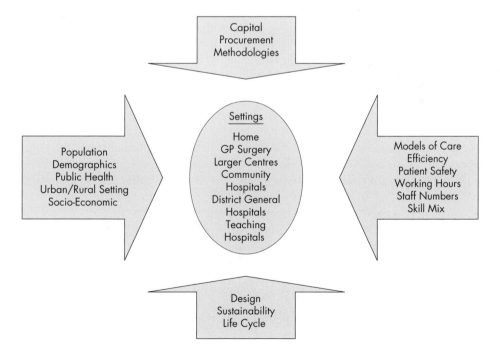

Figure 2.1 Key determinants in planning healthcare environments.

that are currently extant as a means of delivering healthcare facilities. Section 2.6 includes a discussion of the **settings for healthcare** delivery, and Section 2.7 gives an overview of the key **supply** side determinants. Section 2.8 gives an overview of the key determinants for **demand** in healthcare, and finally Section 2.9 ends with a discussion on **design** issues.

These aspects have been discussed in this chapter to give an overview of the factors that need to be considered when planning new healthcare facilities. These factors will change as policy develops and changes.

2.4 Policy developments since 1997

As noted above, since the creation of the NHS, policy has shifted and changed according to the manifesto commitments of successive Governments and the prevailing economic climate. The Hospital Plan of 1962 had far-reaching ambitions, which were derailed by the economic conditions of the time and the macro-economic conditions of the world as a whole during the 1970s and early 1980s.

2.4.1 The NHS Plan, 2000

In 2000, the Government launched the NHS Plan (DoH, 2000c). This was seen as the most significant statement of intent to the development of the NHS, probably since the Hospital Plan of 1962. The plan was designed to be a 10-year strategy of investment in NHS services across the Board and in the development of the workforce and the physical environment. It is difficult to convey the overall feeling of excitement and optimism generated by the release of the plan and the level of support for it across professional bodies, both clinical and nonclinical – as demonstrated by the organisations endorsing the document.

The plan envisaged an unparalleled increase in the revenue allocation for the NHS, with the aim of moving health spending towards the EU average. Significantly, the plan also identified firmly that the NHS would continue to be free at the point of need.

With regard to the healthcare environment, the NHS Plan called for the following:

- £1bn of investment in primary and community care;
- the creation of 500 'one-stop shops' in primary care, bringing together GP services, diagnostics and social care services – which, it could be argued, echoed the call for the development of polyclinics in Lord Darzi's later report into the redevelopment of services in London;
- 100 new hospital projects to be completed or underway by 2010;
- Expansion in diagnostic and treatment services across the NHS.

The development of new acute hospital facilities would be delivered through a combination of traditional public capital funding – which was to be increased – and through the Government's Private Finance Initiative (PFI), which is discussed in Section 2.5.2. The investment in primary and community care would come through a new public–private partnership (PPP) known as the NHS Local Improvement Finance Trust (NHS LIFT), which is also discussed further in the text.

The expansion of diagnostic and treatment services (increased numbers of computerised tomography (CT), magnetic resonance imaging (MRI) scanners and linear accelerators for radiotherapy) was mainly undertaken by one-off programmes of ring-fenced capital from the Department of Health and the National Lottery 'New Opportunities Fund'.

The NHS plan also set exacting targets for expansion of capacity through employing more staff in the NHS (as well as providing the physical environment for them). Through this, patients and the public were guaranteed reduced waiting times and better access to services. Key pledges in the NHS Plan for improving access were as follows:

- All patients should be able to secure appointments with a GP within 48 h.
- By 2004, patients should not be waiting for longer than 4 h in an A&E department before being admitted, treated or discharged.
- The maximum wait for inpatient treatment would be reduced to 6 months by 2005 and to 3 months by 2008. The maximum waiting time in 2000 was 18 months.

This key piece of policy had a dramatic effect on planning healthcare environments: from a position where major healthcare developments were unusual, they quickly gathered in scope, scale, pace and incidence. In the early 1990s there had been perhaps less than a dozen major projects (over £25m). By early 2009, more than 75 have been delivered and are operational and a further 25 are under construction.

2.4.2 Delivering the NHS Plan, 2002

Two years after the NHS Plan, an update on progress was published (DoH, 2002). The NHS Plan had proposed a change in the sometimes difficult relationship between the public and private healthcare provision sectors. Delivering the NHS Plan envisaged that some of the diagnostic and treatment centres proposed in the plan would be delivered by a new form of partnership between the public sector (as a commissioner of services) and the private sector (as a provider).

The Independent Sector Treatment Centre (ISTC) programme was also launched to provide additional capacity provided by private-sector companies for elective procedures. Essentially, new facilities would be provided that would create further NHS capacity for simpler, high-volume procedures to drive down waiting times and waiting lists.

2.4.3 The NHS Improvement Plan, 2004

The year 2004 saw a further update on the progress of the NHS Plan (DoH, 2004). This developed several of the themes in the NHS Plan and confirmed most of which had gone before. The Improvement Plan also focussed on the following aspects:

- delivering improvements in the general patient experience and how these could be measured through regular surveys of patient opinions;
- monitoring of standards through increased inspection of healthcare facilities (both public and private) through the Healthcare Commission, with an increased drive to ensure that hospitals were clean and that hospital-acquired infection rates were decreased;
- introducing a system of payment for providers known as payment by results (PbR), which would set standard national payment rates for particular procedures. This would have the impact of meaning that providers would no longer compete on the basis of cost, but could focus on quality and access times;
- reducing waiting times for treatment from referral by a GP to a maximum of 18 weeks by 2008.

The final key pledge was a slight amendment on earlier policy statements in that it meant that the entire patient care pathway – including any diagnostic tests necessary to inform treatment – would need to be undertaken in the 18-week target that was subtly different from earlier pledges.

2.4.4 Our health, our care, our say: a new direction for community services, 2006

By the mid-2000s, it was becoming clear that although the NHS Plan (DoH, 2000c) and subsequent policy initiatives had looked at all aspects of the NHS in the drive to achieve a more responsive public service with decreased waiting times and enhanced capacity, which went hand in hand with greatly increased resources, the pace of reform in primary and community care was seen by some to feed further impetus.

Our health, our care, our say (DoH, 2006a) confirmed once more that the direction of policy travel was for the development of primary and community-based services. This had been repeated by successive governments since the creation of the NHS, but the main focus in the minds of the public (and politicians) had been about acute hospitals. Our health, our care, our say sought to redress this balance by focusing on primary and community care and encouraging greater cooperation and joint working with social services that were managed and operated by local authorities.

2.4.5 Our health, our care, our community, 2006

In 2006, after an extensive public consultation exercise of attitudes towards health services and their preferred locations, the government launched Our health, our care, our community – investing in the future of community hospitals and services (DoH, 2006b). This set out a clear commitment to delivering healthcare, where clinically appropriate, away from the hospital setting and into locations closer to where people lived and worked. There is a further discussion about this in the settings for healthcare section.

2.4.6 Healthcare for London, 2007

Although focused on London, Lord Darzi's report on improving healthcare in the capital (NHS London, 2007) reflects many of the debates in other health systems, not just in the UK but internationally. It is therefore an important policy development that does need to be considered when planning developments in healthcare facilities.

This was the third major review of the health system in London since 1992, following the Tomlinson Report in 1992 and the Turnberg Report in 1998. Both of these reviews were clear that primary and community care services needed to be improved in the capital and one of the mechanisms for investment in these areas was the rationalisation of acute and specialist services in the hospital sector. Much of the investment in the healthcare infrastructure in London was focused on the delivery of the changes in the acute sector, such as the rationalisation of services between Guy's and St. Thomas Hospitals, proposed by Tomlinson and confirmed by Turnberg. Healthcare for London identified eight key areas for improvement.

- Improving health generally: London has higher rates of childhood obesity than the rest of England. It also has 57% of England's cases of AIDS, has specific challenges in relation to mental health and suicides (being the commonest cause of death for men under 35) and has a highly diverse population with over 300 languages spoken across 90 different ethnic groups – many of whom have specific health needs.
- Patient expectations: The degree of satisfaction with the NHS expressed by the people in London is lower than the average for similar surveys undertaken across England (27% dissatisfied with the running of the NHS in London compared with 18% nationally).
- Significant health inequalities: Life expectancy decreases dramatically from central London to the east end of London – with each tube stop on the Jubilee Line representing 1 year of life expectancy lost. This masks significant variations but the overall message is a powerful one.
- Deliver care in different settings: The report notes that in the US, 90% of outpatient attendances were held in hospitals in 1981, but this had reduced to 50% by 2003 with an increasing number being undertaken in out-of-hospital settings. It is not clear whether this is a like-for-like comparison, given the different nature of the healthcare systems and the fact that one does not provide universal coverage, but it is again an interesting statistics. It was also clear that utilisation of urgent care is also higher in London than the average for the rest of the country.

- Specialised care: The concentration of high-tech equipments and resources is done in one place where there are enough trained staff to use the equipment and the volumes justify the cost of such facilities. However, generally London has one of the smallest catchment populations per hospital in England.
- Be at the cutting edge: There should be better and closer cooperation between hospitals and universities. This need has already led to the establishment of Imperial College Healthcare NHS Trust – a single academic medical centre encompassing Charing Cross Hospital, Hammersmith Hospital, St. Mary's Hospital and the medical faculty of Imperial College London. At least two others, which focus on the medical schools of Kings College London and Queen Mary and Westfield College, are being established.
- Using workforce and buildings effectively: Labour productivity is seen to be lower in London than the average in England, and the disposition and operation of the estate have been identified as areas requiring improvement.
- Better value for money: The average length of stay in London was higher in 2004–2005 than elsewhere, which could lead to 800 000 bed days equating to 2000 beds or a potential saving of £200m – of course, all of these beds would have to be at one place to generate substantial savings. It is more likely that they are dispersed across all the major hospitals in London with the attendant difficulty of removing whole wards or buildings, thereby realising the maximum savings in fixed costs.

Lord Darzi proposed that one way of providing services that are more accessible to patients in London and providing a range of different care services – particularly focusing on the management of long-term conditions – would be to develop polyclinics that would have a large range of high-quality community services with extended opening hours. The report envisaged the polyclinic having the following range of services:

- GP and community services, open 12 h/day;
- outpatient appointments (including ante/postnatal) open 12 h/day;
- minor procedures open 12 h/day;
- urgent care, available between 12 and 24 h/day;
- diagnostics available 18–24 h/day;
- health information and healthy living facilities open 18–24 h/day;
- long-term condition management available 12 h/day;
- pharmacy open 18–24 h/day;
- opticians, dentists etc. available 12 h/day.

The polyclinic would serve a population of about 50 000 and would also have co-located local authority services and leisure facilities where appropriate.

No one solution was set out. It was felt that these facilities could be configured in a number of different ways to meet the needs of the local population – from hub and spoke models through to models that are co-located. However, the debate on change in London's healthcare provision has centred on the single site polyclinic model.

This front line of care would be complemented by local hospitals serving a population of 200–250 000 that would have an A&E, diagnostics, an obstetric unit, inpatient beds, rehabilitation and a high dependency unit but not an intensive care unit. It would not offer 24-h emergency surgery. Elective centres would be similar to ISTCs and provide high-throughput elective surgery with diagnostics and outpatients. Such centres would operate 12 h/day.

The major acute hospital would be open 24/7 and provide all major acute facilities and provide services to a population of 0.5–1m. Those centres that would offer specialist services such as level 1 trauma and transplantation would serve a catchment of up to 5m population. NHS London, the Strategic Health Authority for London, has recently announced a proposed pattern of level 1 trauma centres, hyperacute stroke units and stroke units that has received much less attention (and heat) than the polyclinic proposals, despite, in some ways, being more far reaching.

2.4.7 High-quality care for all, 2008

Following on from his review of health services in London, Lord Darzi was asked to undertake a national review of the NHS and identify its way forward over the next 10 years.

The review was completed in early 2008 and the final report High-quality care for all (DoH, 2008) was published in June 2008 – just in advance of the celebrations for the 60th anniversary of the NHS. The report responded to, and was accompanied by, documents provided by each SHA that set out the local vision for how the NHS could be improved.

The most surprising feature of Lord Darzi's report was that it eschewed the usual tendency for reviews of this nature to be prescriptive, top-down, target-setting exercises. Instead, the review posited that the previous 10 years of development in the NHS had been the drive for additional capacity (staff and buildings) through the dramatic increases in resource that had been provided. The review reflected that the NHS now needed to commit to a similar drive in the quality of services that were to be provided.

The key messages in the report were as follows:

- Providing high-quality care for patients and the public through providing people with a greater degree of choice and personalisation of their care. It also highlighted the roles of Primary Care Trusts (PCTs) and local authorities in playing a role in health promotion and preventing illness. This tied in with the new legal rights and responsibilities that local authorities have in producing joint planning for health and social care.
- Developing a new NHS constitution to set out what the public can expect from the NHS and the rights and responsibilities of the public within that.

- Putting quality at the heart of the NHS through the publication of more information about healthcare services and setting independent quality standards for care. In particular, funding for NHS hospitals will, for the first time, include an element that reflects patients' assessments of their experience.
- Providing greater freedom for the NHS to innovate and manage its services at local level.

Although viewed in one way, it is difficult to immediately see what the impact on NHS infrastructure will be from the Darzi Report. It is, however, clear that good-quality environments can have an impact on the satisfaction reported by patients in their experiences.

2.5 Capital procurement methodologies and NHS organisations

As noted in earlier sections, there are several procurement methodologies extant in early 2009 that can be brought to bear in the delivery of healthcare facilities and these are discussed in the following subsections. This section also deals with how each type of NHS organisation can deploy these methodologies.

2.5.1 Overall capital investment in the NHS
The amount of available capital has increased from around £2bn in 1991–1992 to £5.5bn in 2007–2008 after being reduced towards £1.1bn in the mid-1990s (Figure 2.2).

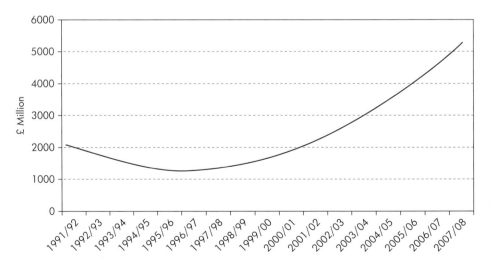

Figure 2.2 Overall Capital Investment in the NHS 1991–1992 to 2007–2008 (DoH, 2007). Crown©: reproduced by permission Department of Health.

It is clear that the PFI pipeline has slowed over the last few years. The main driver behind this was, initially, the deficit incurred in the NHS in 2005–2006, which was £500m, or less than 1% of the total revenue allocation and the impression – rightly or wrongly – that the NHS was building significant capacity in the acute sector (through the capital programme as a whole and the ISTC programme) that was unsustainable given the greater policy thrust towards the development of capacity in the primary and community care sector and the shift of activity – where clinically appropriate – from the acute sector.

The trend over the past 10–15 years has been for decreasing amounts of capital to be held at DoH and more delegated further down into the system. Strategic health authorities held significant amounts of public capital for allocation within their regions. This was changed in 2006 during the last intermediate tier reorganisation, and there is now a significant difference between different types of organisations as to how they access capital for major investment programmes and projects.

2.5.2 The Private Finance Initiative

In 1992, the Government launched the PFI, but it was not until 1995 (after CIM was launched) that the NHS engaged meaningfully and on a significant scale in delivering major hospital developments through PFI.

The tenet at the time was for an unrestricted access to the growing PFI market ('Let a thousand flowers bloom', which is an erroneous quote from Chairman Mao Zedong and attributed anecdotally to the then Chancellor of the Exchequer, Kenneth Clarke), and the extent of the demand for capital in the NHS became apparent as a relatively small number of PFI developers chased an increasing number of major and minor projects.

The PFI meant that the private sector would design, build, finance and operate (for nonclinical services) new hospital facilities. These hospitals would be owned by the private sector and leased back to the NHS on a long-term basis – typically 30+ years. At the end of the contract period, the NHS hospital would have the option of tendering for a further period for private-sector companies to operate the facilities or to take back the hospital under public ownership once the cost of construction had been recouped by the private sector through the long lease–style arrangement.

The advantages of the PFI were seen by successive governments to be as follows:

- Access to another source of finance – the capital markets – would be provided, essentially creating another source of supply against which to meet the burgeoning demand for capital.
- Given that the risks of ownership would lie with the private-sector developers, the assets themselves would not be counted on the balance sheet of the NHS Trust and therefore would not count against government's borrowing targets. This was a particularly important issue for the Conservative government and

the Labour government that would succeed it, as borrowing targets across EU member countries were restricted to potentially allow access to the single European currency.

- The risk of delivering the asset to time and to an agreed cost would be transferred to the private sector – if they got it wrong, they bore the brunt and the public sector would have no obligation to meet any additional costs caused by time and cost overruns. This was particularly important in the light of the various reports published by the NAO and PAC into the delivery of major infrastructure programmes by the public sector.
- The condition of the asset at the end of the contract term would be the same as the time at which it was originally completed – as new – as the maintenance and life-cycling of the asset would be guaranteed under the contract. This was considered important as it meant that the public-sector asset base would be effectively maintained for the first time – maintenance and life-cycle repair budgets have usually been the first areas for reduction in difficult economic times.

After the initial surge in active projects, it was clear that the PFI was struggling in health. The number of schemes being progressed was stretching the ability of the market to resource them effectively. Lack of progress at the national level on reaching at an agreement on standard contract terms meant that many schemes made painfully slow progress but with increasing capital costs and the significant costs to both the private and public sector of the people needed to plan, design and draw up the contractual provisions.

It was not until late 1997 and the support of the new Labour government for the PFI that the first tranche of hospitals signed contracts with private-sector developers. However, this was only after the passing of several pieces of primary legislation that allowed NHS Trusts to have the legal powers to enter into such long-term contracts. At the same time, a scheme of national prioritisation was introduced that reduced the numbers of schemes in procurement, but meant that resources could be focused more effectively on a smaller number of projects (although still fairly significant in value terms). A particularly helpful overview of the history of PFI in the NHS, with a specific focus on accounting, has been developed for Chartered Institute of Management Accountants.

From 1997 till early 2009, 76 NHS PFI schemes worth £5.3bn are operational, 25 schemes worth £5.6bn are under construction, just 1 is currently under negotiation and 9 are yet to gain approval to seek private-sector partners. This is at the time of writing, but is constantly changing as schemes are completed.

2.5.3 NHS LIFT

NHS LIFT was to be developed by an entity that was itself a PPP. A national company – Partnerships for Health (PfH) – was established, which was a joint venture between the DoH and Partnerships UK (PUK). PUK was also a PPP joint venture between HM Treasury and private-sector investors. LIFT was launched formally in 2001 after the commitments made in the NHS Plan of 2000 for investment in primary and community care.

LIFT has some features in common with PFI – essentially it is a vehicle for delivering healthcare facilities in primary and community care that are built, designed, funded and operated by the private sector. The private sector also undertakes the maintenance of the facility.

At the local level, LIFT is a joint venture between a private-sector consortium (comprising building contractors, maintenance contractors and investors), who hold 60% of the shares in the LIFT company (LIFTCo), the local NHS (one or more PCTs), which holds 20% of the shares, and the national joint venture, Partnerships for Health who own the other 20%.

In 2007, PUK sold its interest in PfH to DoH that is now the sole shareholder in the national LIFT company, which is now known as Community Health Partnerships. There are currently 47 LIFTCos established, covering around half of the PCTs in England.

According to the Community Health Partnerships website (seehttp://www.communityhealthpartnerships.co.uk), there has been £1.5bn invested in LIFT, with 220 buildings operational or under construction at the time of writing. The current drive in LIFT is to widen the participation of public-sector bodies in LIFT, essentially to widen the involvement in service delivery within LIFT buildings to include non-NHS service providers. The intention is to provide centres that include NHS and local authority services so that patients and the public are accessing public-sector services in one place, rather than an array of different facilities in an array of different locations.

This reflects the drive towards one-stop shops, which was a feature of the original concept of LIFT but was an exception rather than the rule in early LIFT buildings – mainly because bringing together different public-sector bodies to one place, with their different approval requirements and funding constraints, is a difficult obstacle to clear.

The level of non-NHS involvement is a mixed picture across England but appears to be most consistent in the north-west and north-east – possibly because of the traditional close working across the public sector in those areas and the relatively low turnover of senior staff in those areas.

2.5.4 ProCure 21

The other instrument in the facility procurement armoury of the NHS is ProCure 21 (P21). This was launched by the DoH in 2001 and was a direct response to several reports into government procurement (not just on the NHS) that sought to develop a more collaborative and partnering approach between the private sector and government in the successful and predictable delivery of major infrastructure programmes and projects.

ProCure 21 set out to establish a framework agreement at the national level of a number of principal supply chain partners (PSCPs). These supply chains would comprise building contractors, design practises and other relevant consultants. An NHS Trust or PCT that wished to undertake a publicly funded project (P21 is limited to publicly funded projects) would select a shortlist of PSCPs from the national framework and seek proposals from each one, setting out the approach the PSCPs would take to working with the NHS, plus some outline proposals for how they would go about this and the experience and expertise they would bring to bear.

The NHS would then select one P21 partner with which it would develop the investment proposal. In many circumstances, the P21 partner would develop the business cases for the project and then move through to actually deliver the new facility. The overall objective is to ensure that there is realism and deliverability in the proposals set out, using the experience of the contractor to ensure that the preferred way forward is actually deliverable.

Once the project is fully designed, the NHS body and the P21 partner agree a guaranteed maximum price (GMP) for which the PSCP will agree contractually to deliver the facility to time, cost and an agreed level of quality.

The record for P21 to date appears impressive. The P21 website (see http://www.nhs-procure21.gov.uk/content/performance.asp) identifies performance metrics summarised as follows:

- Usually more than 80% were delivered on time.
- More than 90% were delivered to cost.

The other intention was that an NHS body might develop a long-term relationship with one or two PSCPs and use them for a pipeline of projects over time. The PSCPs would understand the NHS organisation and the sites on which they were delivering the schemes – this would lead to greater standardisation in the design process and allow the PSCP to drive continuous improvement in project delivery. A DoH central team collects the data and ensures that there is benchmarking information available to the NHS to ensure that the costs being proposed by the local PSCP are consistent with those being offered elsewhere for similar types of projects.

This can only really be demonstrated where NHS bodies decide to use P21 partners for a wide variety of projects over time. There is some anecdotal evidence that P21 is only really effective on projects that are over £5m in value to justify the overhead costs that are built into the PSCP costs.

P21 was put in place for a 7-year period, and the framework expires in 2010 (after being extended for 2 years). The DoH is currently consulting upon how a replacement framework will be established to replace it.

2.5.5 NHS Foundation Trusts

NHS Foundation Trusts (or FTs) are able to retain internally generated capital from the depreciation on their assets. FTs have a flexible accounting regime that allows them to treat capital and revenue as a single resource. FTs can also retain any surpluses generated from their day-to-day activities. Both of these can be then used and accumulated over a period of years for internal investment – in facilities, equipment or the workforce. By February 2009, 114 FTs had been authorised to operate.

NHS FTs also have the ability to borrow from a DoH central resource known as the NHS Foundation Trust Financing Facility (FTFF). However, an FT is only allowed to borrow up to a cumulative total known as the prudential borrowing limit (PBL). The PBL is set and reset from time to time depending on the relative financial strength of the organisation. The loan taken from the FTFF is then paid back at a quasi-commercial interest rate (usually set as the National Loans Rate plus 1%) over an agreed period linked to the lifetime of the asset. The PBL essentially sets the limit for borrowing, which is sustainable by the projected income and expenditure of the organisation.

FTs do have the ability to borrow from commercial organisations such as banks, but any borrowings also count against the PBL in the same way as borrowing from the FTFF. It is likely that commercial borrowing will be at a higher interest rate than from the FTFF and, therefore, will be less attractive overall.

FTs can enter into PFI contracts but will need to demonstrate to the regulator, Monitor, that they can still meet their financial duties under the terms of their licence to operate as an FT. As non-FTs, FTs will also need a 'Deed of Safeguard' from DoH to enter into the PFI contract, and this approval requires them to seek approval for their final full business case for the PFI transaction.

2.5.6 NHS Trusts

Until 2006, NHS Trusts received an allocation of capital (known as 'operational capital'). This was allocated on the basis of the depreciation charge on the assets that the Trust owned and operated from. This was mainly intended to maintain the current asset base and ensure that it was safe from which to operate from staff and patients.

NHS Trusts also had the opportunity to bid for strategic capital. This was a fund of capital held by Strategic Health Authorities, the quantum of which was decided usually on a capitation basis (i.e. the size and demography of the population within that region). Individual Trusts and PCTs were then able to bid for funds from this programme for specific schemes.

In practice, for major schemes (usually defined as being above £25m) Trusts and PCTs would use either the PFI or LIFT route. There have been few schemes over £25m over the past 10 years that have been funded from public capital.

In 2006, the amount of strategic capital was reduced dramatically and, until 2008, was only made available – generally – to PCTs for their developments. NHS Trusts were to retain the depreciation charge on assets employed in full. The objective here was to align the financial regime of NHS Trusts with FTs as increasing number of Trusts became FTs.

2.5.7 PCTs

PCTs were also provided with operational capital (on the same basis as NHS Trusts) and could also bid for strategic capital from their local SHAs. In practice, the amounts of operational capital were relatively small, in comparison to that allocated to the acute and mental health Trusts because of the relatively small nature of the PCT-owned asset bases. This system was changed in 2007.

PCTs are now expected to design a capital investment programme that is affordable within expected resources and submit this plan to the SHA. The SHA then collates the plans and submits these to the DoH for capital funding to be allocated. The expectation is that the plans will be funded in full. However, the plans must demonstrate that they can be delivered within the key milestones that the PCT has set out or the PCTs risk losing the capital allocation. PCTs also have access to the LIFT initiative.

2.6 Settings for healthcare

There are a variety of settings for delivery of healthcare:

- in the home;
- in GP surgeries;
- in larger health centres;
- in one-stop shops/joint service centres/polyclinics;
- community hospitals;
- district general hospitals/local acute hospitals;
- Teaching hospitals – particularly for complex and tertiary care.

By early 2009, the NHS had around 1600 buildings identified as hospitals, 29 000 general practitioners (GPs), 18 000 dentists, 27 000 pharmacists, 30 000 hospital consultants and almost 400 000 nurses.

The NHS operates from over 25m m^2 of occupied floor space – more than 5000 standard football pitches. This does not always include GP practises owned by the GP or by a third party.

2.6.1 The home

This is often overlooked in terms of delivery of health and other care. Home care, delivered by a range of practitioners, from GPs to community care staff, to social services and even outreach from the acute sector. Care such as podiatary through to such innovations as simple chemotherapy can be delivered in a home setting.

Most of the care undertaken at home requires little in the way of adaptation to the home environment. However, with increasing innovation in technology, it has become increasingly possible to provide far greater monitoring of patients in their homes, based on a risk assessment of their conditions.

Telemedicine and telecare, discussed in Chapter 9, are established technologies, but the growth in the uptake of these has been slow. As part of the DoH's care outside hospital programme, three national pilots in the application of telecare have been established – in Kent, Newham and Cornwall. These pilot programmes are currently taking part in one of the largest nonrandomised evaluation trials in history to establish clearly whether the programme can deliver the benefits that have been ascribed to it.

It is entirely possible that if these programmes can demonstrate substantial benefits to patient care, then the uptake in such technologies will increase. It is not clear at the moment whether this will allow a substitution of capacity from the primary and community setting or from the acute setting of both, or what the relative costs of these will be. It is not beyond the realms of possibility that there may be some opening up of unmet demand, rather than substitution from other settings.

2.6.2 General practitioner surgery

Although much of the debate in planning healthcare environments has focused on the acute hospital setting, almost 90% of all the patient contacts that happen in the NHS on a daily basis are in primary care.

Increasingly, primary care is being delivered in larger centres that provide the opportunity for co-location and integration with other services. However, there is still a considerable amount of primary care delivered in GP surgeries that may contain one or more GPs, but little else in the way of supporting services. GP surgeries can be configured in a number of different ways and in a variety of different ownership models:

- Traditionally, especially in deeply rural areas and in certain inner-city areas, GP surgeries were delivered from the GPs own house – with one or two rooms turned over to reception and consulting facilities.
- GP surgeries were also delivered in small health centres, which mainly focused on consulting and examination facilities. Some centres had their own dispensing pharmacies attached. These centres could be owned by the GP themselves, by the local PCT or by a third party who would charge the GPs using it at an agreed rental over time (this system is known as third-party development).

One of the key features of facilities that are not owned by the NHS is that they did not have to comply with NHS standard – space or engineering. As long as the local NHS was prepared to pay the rental involved, through reimbursement of the GPs involved, there was also little requirement for approvals beyond the local health economy.

The main advantage to GPs of self-ownership was that as independent contractors, and not employed directly by the NHS, that they could sell on the interest in the property when they retired or moved on and keep the proceeds. Given that GPs did not have access to the NHS pension scheme and the advantageous terms that this offered, this was essentially the GPs pension plan. It would appear that there are moves for the new Care Quality Commission to have inspection rights over all facilities from which care commissioned by the NHS is delivered; this includes private hospitals, and there appear to be moves to extend this to premises owned by GPs and third-party developers. This will, for the first time, create a level playing field in primary care provision between GP-owned facilities and those that have been funded by the NHS and that have to meet NHS standards.

2.6.3 Larger health centres

Since the 1960s, there has been an increasing trend for the development of larger health centres that contain more patient services under one roof. These larger centres might include teams of GPs, community services teams, podiatry, rehabilitation and acute hospital outreach clinics. Over the past 10–15 years, there has been an increasing trend to offer minor treatments and, in some cases, procedures that would have only been undertaken in a hospital setting such as vasectomies. There has also been a marked increase in the amount of diagnostic imaging and especially obstetric ultrasound being undertaken in this setting.

2.6.4 One-stop shops/polyclinics

More recently, and following the publication of the NHS Plan in 2000 (DoH, 2000c) and subsequent policy initiatives, there has been a trend towards ever larger facilities in the primary care setting – the NHS Plan defined them as one-stop shops where a variety of health and social care services could be provided. There was also the opportunity to provide other services, such as benefit payment offices, jobcentres and post offices in these areas.

The growth in these types of facilities has been accelerated by the government policy in developing 'joined-up' services across the health and social care boundaries, with the expectation that co-location and even integration of services in one building could provide significant benefits to patients as they would have a single point of contact for public-sector services in one place.

The concept of the polyclinic is not a new one: one of the first to be badged in this way was opened in Hove in 1993. Depending on which definition is applied, many of the larger health centres and one-stop centres could easily be described as polyclinics. As noted above, there has been considerable debate about the implementation of the polyclinic model in London (NHS London, 2007), but the concept has been discussed in the NHS for quite some time – even featuring in Acute Futures (Harrison and Prentice, 1996) from 1996.

2.6.5 Community hospitals

With the publication of Our Health, Our Care, Our Say (DoH, 2006a) and Our Health, Our Care, Our Community: Investing in the Future of Community Hospitals and Services (DoH, 2006b), the community hospital was given increased prominence in the healthcare system. The revitalisation of these facilities – which

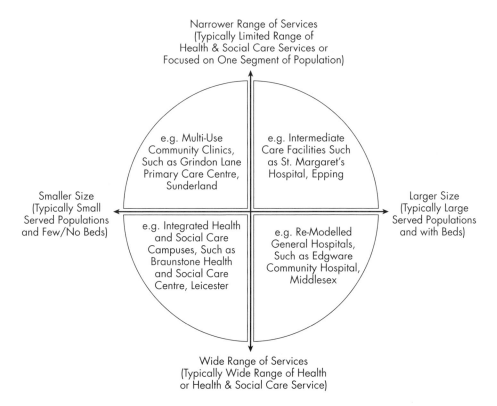

Figure 2.3   The continuum of services to be provided in different types of community hospitals, (DoH, 2006b). Crown©: reproduced by permission Department of Health.

had seemed to be under threat following the centralisation of services over the past 20 years – was central to this.

Community hospitals as currently defined can provide quite a broad range of services as shown in Figure 2.3.

The Department of Health (DoH, 2006b) announced that £750m would be available from the overall capital allocation to support the development of community hospitals and services over a 5-year period.

Our Health, Our Care, Our Say: Investing in the Future of Community Hospitals and Services (DoH, 2006b) identified 10 key design principles (design here referring to the design of services within the building) for community hospital–type facilities. They should be:

- locally led;
- redesign patient pathways – in particular to ensure that care can be transferred from the hospital setting or not put there in the first pace if this is a new service;

- anticipate future needs;
- adopt new technologies;
- plan across primary and secondary care;
- affordable to the whole health economy;
- integrated with other health and social care services;
- engage with staff to ensure that their potential is recognised;
- enable any transitions staff may have to make if the new facility involves transfers of staff from other organisations or settings;
- be of high quality.

There are also some significant successes in bringing various services together under one roof. The LIFT and community hospital programme has promoted this, as has the joint service centre initiative that is supported by the Department for Communities and Local Government.

2.6.6 District general hospitals

District general hospitals have been one of the main areas for investment, as noted earlier. However, this sector has not been immune from the various changes in provision being proposed. There have been many changes in the pattern and location of DGH services since their inception. In some cases, the hospital investment programme has extended, replaced and rationalised DGH services. There are still a significant number of changes in DGH provision that are being examined:

- District general hospitals becoming local acute hospitals that provide a service more focussed on elective services and that may have a limited emergency or urgent care service – with full A&E services available 24/7, being provided in a nearby hospital.
- To balance this, some DGHs are becoming more specialised or entering into clinical networks with other local providers.

2.7 Supply-side considerations

In planning healthcare environments, there are many determinants and inputs that have to be considered to reach an optimum balance in meeting demand-side requirements and pressures. The supply-side elements will change depending on the nature of the healthcare setting (as discussed earlier), but many of the principles are similar.

As noted previously, the issue of the bed in NHS planning is a key one and the one that generates the most interest and excitement as developments or changes to health systems are planned and discussed. The trends that are experienced are useful indicators in particular areas (beds, accident and emergency attendances, outpatient attendances and imaging modalities) for future provision.

2.7.1 Beds

There are several classifications of beds in the NHS:

- general and acute
- day
- intermediate
- geriatric
- mental health
- learning disability.

Generally speaking, over the past 10 years, the first three categories have increased in the number and the final three decreased, as shown in Figure 2.4. General and acute beds can be one of the most contentious areas for planning purposes, with maternity coming a close second.

The traditional methodology for calculating beds has been to identify current activity and then apply a level of growth agreed with commissioners over a planning period – usually 3–5 years, but with a 10–15 year horizon required for major developments. It is then usual to discount this by application of efficiency measures

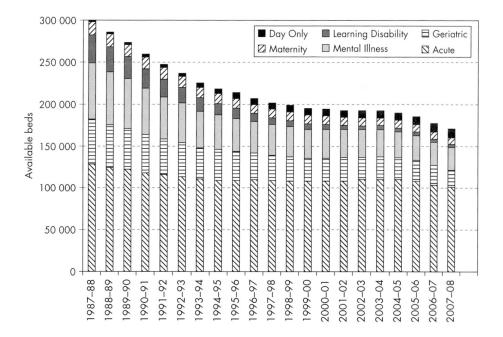

Figure 2.4   Time series of daily available beds in the NHS, 1987–1988 to 2007–2008.

Source: DH Hospital Activity Statistics (see http://www.performance.doh.gov.uk/hospitalactivity/data_requests/index.htm).

such as reducing length of stay or admission avoidance measures. An alternative methodology is to look at the length of stay cohorts and identify a 'cut-off' point where patients in acute hospital settings can be considered to be medically stable and therefore require less acute care – such as transfer to an intermediate care setting or to a home environment if the requisite support care is in place to allow this.

This latter point has been one of the key issues in hospital facility planning – the level of 'delayed transfers of care', which mean that a proportion of beds are taken up by patients who do not necessarily need an acute hospital bed but are not able to be transferred to a more appropriate setting. This is one of the major contributory factors to the usage of beds in London.

2.7.2 A&E

As can be seen from Figure 2.5, A&E attendances – particularly new attendances – have risen dramatically in England over the period from 2003 to 2004. There is evidence to suggest that this increase could be attributable to the equally dramatic performance improvements in A&E that were required under the NHS Plan – that waiting times in A&E before treatment or admission should be no longer than 4 h. The hospital system in England is running at around 98% compliance to this standard, so it is possible that faced with a 2-day wait to see a GP (another national standard) or a visit to A&E where you will be seen and treated within 4 h, individuals are voting with their feet and going to A&E, which traditionally was the place of final resort.

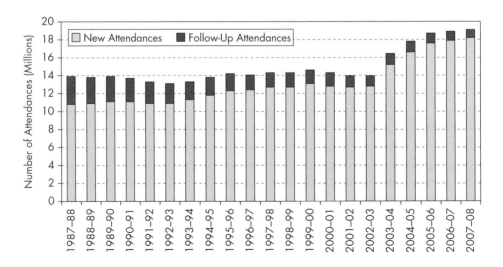

Figure 2.5 Time series of Accident & Emergency Department attendances in the NHS, 1987–1988 to 2007–2008.

Source: DH Hospital Activity Statistics (see http://www.performance.doh.gov.uk/hospitalactivity/data_requests/index.htm).

The level of admissions through A&E has also risen in proportion. There is also seasonal variation associated with these figures, but over the last few years an interesting phenomenon has occurred – that A&E activity in summer has overtaken activity in winter; in particular, the busiest traditional times for A&E activity are just after Christmas (when people present after feeling ill after Christmas but put off a visit to A&E until Christmas is over) and late January/early February during the traditional coldest times of the year when A&E attendances are dominated by falls and fractures (especially in wintry weather) and upper respiratory tract infections (caused by cold and wet weather).

2.7.3 Outpatients

Equally, as seen in Figures 2.5 and 2.6, the number of follow-up attendances in A&E and at OPD has fallen over the period, and the direction of travel as set out in DoH (2006a) and DoH (2006b) is to reduce these even further by undertaking these in settings that are closer to the patients home or work environment where it is clinically appropriate to do so.

2.7.4 Imaging

Imaging (X-ray, CT, MRI etc.) has also increased in prominence and in intensity, as demonstrated in Figure 2.7.

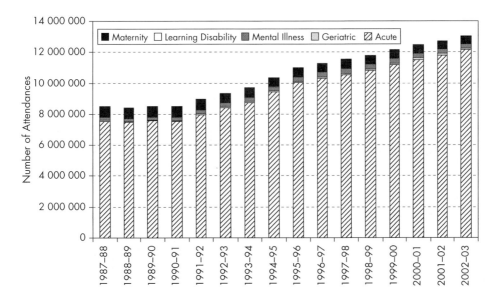

Figure 2.6 Time series of first outpatient attendances in the NHS, 1987–1988 to 2002–2003.

Source: DH Hospital Activity Statistics (see http://www.performance.doh.gov.uk/hospitalactivity/data_requests/ index.htm).

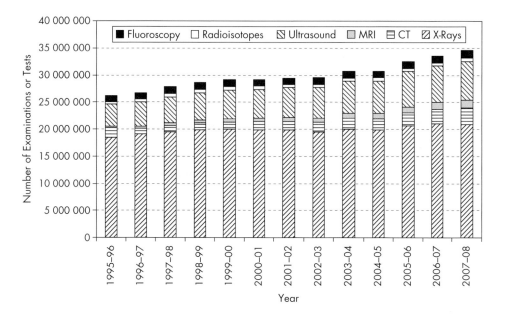

Figure 2.7 Time series of imaging and radiodiagnostic tests, by imaging modality in the NHS, 1995–1996 to 2007–2008.

Source: DH Hospital Activity Statistics (see http://www.performance.doh.gov.uk/hospitalactivity/data_requests/index.htm).

2.7.5 Other factors

As well as the more obvious supply side issues, there are additional factors (performance and wider infrastructure) that must be taken into account:

- Travel times and access: These are important factors not just for patients but also for the emergency transport system (ambulance and paramedics).
- Length of stay: Measures have to be taken to reduce lengths of stay in acute facilities through different models of care within the hospital – for example, by more aggressive models of rehabilitation undertaken at very early stages for patients who have suffered a stroke or serious injury.
- Occupancy: The consultation on the National Beds Inquiry (DoH, 2000a,b) indicated that the optimum level of bed occupancy in an acute setting was no more than 85% to ensure that emergency and elective activity did not impact on each other.
- Scan times and modalities: Although scan times can be improved marginally through better technology, the time required to prepare, scan and remove the patient are broadly similar within modalities and there is little efficiency here.
- Operating procedure times: The procedure times have to be calculated in advance, such as how many patients can be treated in a session, and how

many sessions are possible per day and per week given that some down time for planned preventative maintenance will be required.

- Consulting times: Time required for new and follow-up attendances has to be calculated.
- Working times: The implementation of the European Working Time Directive means that staff should not be working for more than 48 h/week, which has huge implications for the NHS given the traditional culture of long working weeks – particularly for junior doctors.
- The whole system: Acute facilities cannot be considered in isolation from other parts of the system and vice versa. This was highlighted in DoH (2000a), and it important to ensure that the interplay between different parts of the health-care system are seen as a continuum and that capacity is planned accordingly as policies to manage demand and move care closer to home (DoH, 2006a, b) become more consolidated.

A key factor that affects capacity has been the drive to reduce waiting times to 18 weeks from initial diagnosis to treatment. In 2000 there were many people waiting for up to 18 months for treatment; over 250 000 patients were waiting longer than 6 months, which underlines the scale of the organisational task that was to be delivered across the NHS.

Different parts of the supply-side equation are applicable depending on the nature of the setting and also the overall operating hours of the facility proposed.

Also, part of the drive for increased efficiency and productivity (see http://www.productivity.nhs.uk) is to take care of the following:

- increasing day case rates;
- decreasing preoperative admissions;
- managing variations in surgical thresholds;
- Managing variations in emergency admissions;
- managing variations in OP appointments.

The DoH Opportunity Locators (created by the NHS Institute for Innovation and Improvement) provides figures for each PCT in England, allowing commissioners to identify the key areas in their health economies for improvement and to focus on areas of potentially greatest benefit (see http://www.institute.nhs.uk/opportunitylocator/).

2.8 Demand side

As summarised in Figure 2.1, there are a number of key considerations to be examined on the demand side:

- The size of the local population – now and in the future: The Office of National Statistics (ONS) is able to provide details of the population by

sex and age cohorts for PCT and SHA areas (see http://www.statistics.gov.
uk/statbase/product.asp?vlnk=997). The current dataset includes the period
2006–2031. These do need to be read alongside local population projections
(such as planned areas of population growth for housing etc.) so that planned
changes not captured within the ONS dataset can also be considered. An
important factor will be into which age cohort the major changes are likely
to occur as certain age groups are more intensive consumers of health services
than others.

- What is the current health of the local population and where the key health
  challenges are? One particular tool available for commissioner of service has
  been prepared by the Association of Public Health Observatories and hosted
  on the London Health Observatory website. This is known as the Health
  Inequalities Intervention Tool (see http://www.lho.org.uk/NHII/Default.aspx).
  The tool examines the health and morbidity of the local population for a PCT
  area and then allows commissioners to model specific interventions – such as
  increased smoking cessation services – and predicts what impact this will have
  on that population and the mortality within that population. At the moment,
  the tool only examines smoking cessation, interventions to reduce infant
  mortality, treatment with antihypertensives and treatment with statins.
- Joint Strategic Needs Assessment: Commissioning PCTs are charged with
  improving the health of their local populations and there is now a legal
  duty for all public-sector organisations to work together to develop a Joint
  Strategic Needs Assessment (JSNA) to demonstrate how they will each play
  a part to do this. This is a key requirement of the Local Government and
  Public Involvement in Health Act 2007, and the JSNA came into effect on
  1 April 2008. These assessments and the plans that derive from them need
  to be compiled at a population-wide and individual basis (see http://www.
  dh.gov.uk/en/Managingyourorganisation/JointStrategicNeedsAssessment/
  index.htm).
- PCT commissioning plans: These now have to provide a detailed plan for how
  health services will be commissioned and how this commissioning will be undertaken and from whom. This has an important part to play in how providers
  operate and plan their developments to meet commissioner need.

Public health input into the demand side model is extremely important. Also
important is for commissioners to identify robust plans to manage the demand
side:

- How can services be redesigned to reduce the demand into the acute
  environment?
- How can services be redesigned to ensure that delayed transfers of care are
  minimised? – This will need the cooperation among primary care providers,
  social services and carers.

2.9 Design and the physical environment

The physical environment in which care is delivered is important and the NHS has generally achieved much over the past 10 years given the nature and scale of the investment that has been made available – £29bn of capital over the past 10 years (DoH, 2007).

There is little really robust research into the role of the physical environment in health – especially in England. It is a sad reflection that the nature and quality of the research undertaken in England and how this was fed into guidance was once the envy of the world. This research base is now extremely thin, but the quality of available evidence globally is fairly thin.

The Centre for Health Design in America has made some attempts to redress this balance, and it publishes some of the more rigorously researched findings and summaries of the available literature. However, an evidence base that links key determinants to patient outcomes – in the same way as evidence based medicine – has nowhere near the depth of that discipline. For example, Rubin and Owens (1996) undertook a literature search in 1996 that looked at 38 000 studies undertaken in the past 30 years: 48 were selected for containing relevant data and 42 of these demonstrated that some healthcare environment feature was related to a patient outcome.

Given the scale of the health building programme over the last decade, there has been considerable interest in the part that healthcare infrastructure design has had to play in the built environment discipline. However:

> Few recent hospital designs have made a positive contribution to mainstream architecture, and as places may arguably inhibit rather than contribute to the healing process. Contemporary architectural criticism outside a few specialist publications seems to ignore the field, and few design and professional awards are made for health care buildings. Health care buildings design is not perceived as a fashionable design arena for either practice or for schools of architecture. (Francis *et al.*, 1996)

Indeed, the wider design and architectural hierarchy have been generally unmoved by the quality or design exhibited by any of the major healthcare programmes. Whether this is attributable to the actual design quality or to the nature of architectural training and the way in which that works is debateable. There is a wealth of published information that relates to the healthcare environments, and most of it can be found as official guidance on the Department of Health website, in the 'Estates and Facilities Management Knowledge and Information' portal. These are mainly categorised as follows:

- Health-building notes: These provide guidance on how, predominantly, acute hospital departments could be configured on the basis of expert panel input. Those planning healthcare facilities should use these for what they are: guidance and nothing more. They are a helpful starting points to provide a baseline against which to plan a healthcare facility but should not be followed slavishly.

• Health technical memoranda: These provide guidance on technical issues within healthcare facilities as the name suggests.

Project teams: The teams should take great care in identifying what is mandatory, what is guidance and what is 'nice to have' within the extant documentation. In some areas the guidance can reflect good practice, and in other areas it can be outdated and contradictory. The design of healthcare facilities is further discussed in Chapters 3 and 4.

2.10 Conclusion

In the past 10–15 years, the NHS has seen a rapid increase in resource, taking it close to the EU average, with average waiting times dramatically decreased and an equally rapid transformation of the estate. In 2000, over 40% of the estate predated the establishment of the NHS. This is now closer to 15%, and much of what has been delivered is not straight like-for-like replacement but new facilities reflecting new models of care to increase capacity and bolster patient choice. These achievements should not be underestimated.

Planning healthcare environments is a complex exercise, undertaken in a landscape that balances political, policy, design and human choice and preference factors. Whatever else will change in policy terms or through developments in medical, surgical and technological expertise, the challenge of delivering modern, efficient, fit for purpose healthcare facilities will remain.

References

DHSS (1983) *Under-Used and Surplus Property in the National Health Service: Report of an Enquiry into Ways of Identifying Surplus Land and Property in the National Health Service*. London, DHSS. ISBN: 0113208278 (Known as the Ceri Davies Report).

DHSS (1984) *National Health Service Management Inquiry*. London, DHSS. ISBN: 0946539014 (Known as the Griffiths Report).

DoH (1989) *Working for Patients*. London, HMSO.

DoH (2000a) *Shaping the Future NHS: Long Term Planning for Hospitals and Related Services: Consultation Document on the Finding of the National Beds Inquiry*. London, DoH (see http://www.dh.gov.uk/en/Consultations/Closedconsultations/DH_4102910).

DoH (2000b) *Shaping the Future NHS: Long Term Planning for Hospitals and Related Services: Consultation Document on the Finding of the National Beds Inquiry – Supporting Analysis*. London, DoH (see http://www.dh.gov.uk/en/Consultations/Closedconsultations/DH_4102910).

DoH (2000c) *The NHS Plan. A Plan for Investment, a Plan for Reform*. London, HMSO.

DoH (2002) *Delivering the NHS Plan: Next Steps on Investment, Next Steps on Reform*. London, HMSO. ISBN: 0 10155 032 4.

DoH (2004) *The NHS Improvement Plan: Putting People at the Heart of Public Services.* London, HMSO. ISBN: 0-10-162682-7.

DoH (2006a) *Our Health, Our Care, Our Say: A New Direction for Community Services.* London, HMSO. ISBN: 0101673728.

DoH (2006b) *Our Health, Our Care, Our Community: Investing in the Future of Community Hospitals and Services.* London, DoH.

DoH (2007) *Rebuilding the NHS: A New Generation of Healthcare Facilities.* London, DoH (see http://www.dh.gov.uk/en/Publicationsandstatistics/Publications/PublicationsPolicyAndGuidance/DH_075176).

DoH (2008) *High-Quality Care for All: NHS Next Stage Review Final Report.* London, HMSO. ISBN: 978-0-10-174322-8 (Known as the Darzi Report).

Francis S., Glanvilleb R., Noble A., & Scher P. (1999) *50 Years of Ideas in Health Care Buildings.* London, Nuffield Trust. ISBN: 1-902089-20-0.

Harrison A. & Prentice S. (1996) *Acute Futures.* London, King's Fund Publishing. ISBN: 1 85717 092 X.

NHS Executive (1994) *Capital Investment Manual.* London, HMSO. ISBN: 0 11 321776 5. The Capital Investment Manual comprises eight booklets:

- *Overview.* ISBN: 0 11 321718 8.
- *Project Organisation.* ISBN: 0 11 321719 6.
- *Business Case Guide.* ISBN: 0 11 321720 X.
- *Management of Construction Projects.* ISBN: 0 11 321723 4.
- *Private Finance Guide.* ISBN: 0 11 321721 8.
- *IM&T Guidance.* ISBN: 0 11 321722 6.
- *Commissioning a Healthcare Facility.* ISBN: 0 11 321724 2.
- *Post-project Evaluation.* ISBN: 0 11 321775 7.

NHS London (2007) *Healthcare for London. A Framework for Action.* London, NHS London (see http://www.healthcareforlondon.nhs.uk/a-framework-for-action-2/).

Rubin H.R. & Owens A.C. (1996) *An Investigation to Determine Whether the Built Environment Affects Patient Medical Outcomes.* USA, Centre for Health Design.

Plan for Uncertainty: Design for Change

3

Susan Francis

This chapter aims to provide an overview of issues around designing healthcare environments. It discusses some healthcare challenges and how these impact on the physical environment. The chapter also discusses strategies to optimise and future proof design and to support the improvement of the quality of healthcare environments.

3.2 Context

Healthcare in the UK has been the focus of significant capital investment since the launch of the government's NHS Plan (DoH, 2000) that set out a programme of investment and reform, including the building of 100 new hospitals by 2010, 500 new one-stop primary care centres, over 3000 modernised (GP) primary care centres as well as modern IT systems in every hospital and GP surgery. The development and delivery of this investment has taken place in a culture of considerable change in terms of not only procurement but also healthcare policy and delivery.

The conventional shape and delivery of healthcare services is changing: rising patient expectations as consumers of healthcare, the development of the expert patient with access to digital health information and the call for a more holistic approach to health and well-being are impacting on healthcare delivery. Demographic trends show significant growth in 45–75-year age group, calling for a focus on the health needs of older people. The proportion of smaller household sizes is increasing, and with more single-parent families and people living alone, the informal support from carers can no longer be assumed. Advances in medical procedures and the use of information technology for support and information are changing how clinical staff interact with patients, where these services are located and how information is transmitted about the procedures.

Government policy to increase patient choice, develop Foundation Trusts[1], introduce national procurement and payment mechanisms all provide a context that is not always easy to join together at project level. But the general directions are clear: care closer to home (and work) (DoH, 2006); patient-driven care with more clinical standards and fewer targets (DoH, 2005a); the development of managed care pathways rather than episodic care interventions; the NHS as a commissioner of services (DoH, 2005b) with a diversity of (service and infrastructure) providers; and the improvement of partnership working in the supply chains.

The Department of Health White Paper (DoH, 2006) set out a vision to provide people with good-quality social care and NHS services in the communities where they live. NHS services are expected to become more responsive to patient needs and prevent ill health by the promotion of healthy lifestyles. Social care services are also changing to give service users more independence, choice and control. DoH (2005a) explains how the NHS Improvement Plan will be delivered. It describes the major changes underway and how some of the biggest changes will be carried forward for a patient-led health service. Following that, DoH (2005b) describes how the Department of Health will develop commissioning throughout the whole NHS system, with some changes in function for primary care trusts and strategic health authorities.

3.3 Impact on the built environment

The impact of the above-mentioned policy drivers on the built environment for health is significant, with a clear signal to move services out of acute hospitals into community- and home-based settings. Currently, four main settings for health are being explored, as shown in Francis and Granville (2001, Figure 13, p. 51):

- Home: for self-care, management of long-term conditions, with carers and expert patients.
- Healthy neighbourhoods: with health integrated with local services such as social care, education and leisure.

---

[1]NHS Foundation Trusts are a new type of NHS organisation, established as independent, not-for-profit public benefit corporations with accountability to their local communities rather than central government control. The Secretary of State for Health has no powers of direction over them. NHS Foundation Trusts remain firmly part of the NHS and exist to provide and develop health-care services for NHS patients in a way that is consistent with NHS standards and principles – free care, based on need not ability to pay. Clinical activity for private patients is strictly limited. NHS Foundation Trusts have greater freedoms and flexibilities than NHS Trusts in the way they manage their affairs. Further information is available in the Department of health website http://www.dh.gov.uk/en/Healthcare/Secondarycare/NHSfoundationtrust/DH_4062852.

- Local hospital: basic diagnostic interventions and treatments in community settings that may be dispersed.
- Centralised campus hospital: for specialist services bringing hospital, biomedical research and teaching together.

Care at home can be both convenient for patients and effective. It can help in avoiding hospital admissions, enabling people to live more independently but with appropriate support. Technology such as monitors and miniature diagnostic kits can provide important information for self-care, enabling patients to call for clinical support as and when they need it. The potential for email, web and telephone consultations is now being used by some GP practices to improve access to care without needing person-to-person consultations.

The recognition that health and well-being are related to the quality of the environment has been understood for sometime. Influential research in the USA and the UK has highlighted the potential of the design of the built environment to support healing and impact on patient health outcomes. The notion of the therapeutic environment is commonly accepted now: whilst 10 years ago we enquired whether the environment could impact on healing, this question has now become more precise, that is, just how does it impact? (Lawson and Phiri, 2003). There are now almost a 1000 international scientific studies supporting this notion, using reliable methods to determine the results (Ulrich et al., 2008). Typical attributes for a therapeutic environment include daylight, views, connection to nature, inclusion of visual and performing art. Some of these are illustrated in Figure 3.1.

Some of the first 'public health' interventions were undertaken in the late 19th century to improve general sanitation and air quality with the then rapidly developing urban centres. Recently, there has been significant interest in drawing together the links between well-being and public health issues such as obesity to fully understand the potential for the environment to encourage and support healthy lifestyles. This means thinking about not only health buildings but also the spaces in between. Commission for Architecture and the Built Environment (CABE) recently collaborated with the National Heart Foundation and Living Streets to publish a report highlighting the need for creating and enhancing places for healthy, active lives (Cavill, 2007). This report covers issues ranging from strategic and urban planning, through walking and cycling, to urban green space and building design. The impact of the environment on mental well-being is highlighted in a Foresight programme that advises the government on how to achieve the best possible mental development and mental well-being for everyone in the UK in the future. This also includes considerations about the design of the physical environment (Foresight, 2008).

Travel plans, sports and leisure facilities, education, health and social care need to be considered alongside one another to create a more joined-up approach to community planning and support. Some of the new projects developed in community settings have sought to generate these connections, and do so in neighbourhoods

Figure 3.1 Thelma Golding Health Centre in Hounslow (now known as the Heart of Hounslow Centre for Health), London, UK.

that are being even more widely regenerated both physically and economically (NHS Confederation, 2006). One of such projects is illustrated in Figure 3.2.

Making better use of local care centres, including community hospitals, and health and resource centres, will be key to bridging the gap between hospital and home. The needs of each community facility must relate to the health needs of the population, and some community facilities may not even require inpatient services and bedroom accommodation. Over the past 10 years, community hospitals have shifted from slow-stream convalescent facilities to becoming predominantly ambulatory local care centres with a wide range of services, often in partnership with other organisations. They have the potential to become fully integrated with other services to provide diagnostic services and treatments in accessible and convenient

Figure 3.2   Luton drop in centre, Luton, UK.

locations close to where people live and work. These trends have been given greater focus by the report by Lord Darzi entitled High Quality Care for All: NHS Next Stage Review Final Report (DoH, 2008), which sets out a vision for an NHS with quality at its heart.

The potential for new ways of working, and extensive use of technology for diagnostics and booking is considerable, with close working connections to the local population that offer continuity of care. The role of the hospital must become the place of last resort for the complex and sophisticated interventions requiring specialist skills and equipment. The location of these streamlined services in convenient town centres with easy links to biomedical research and teaching facilities will become paramount. But they can also play a key part in urban regeneration, being significant public buildings and landmarks. Bringing these services together and making the connections spatially to link to pedestrian routes and civic places are all part of making health an active contributor to the public realm.

3.4 Optimising design

Designing new buildings offers a unique opportunity to provide efficient, safe, quality environments to meet the needs of a modernised health service. New buildings can be a catalyst for change affecting behaviour and questioning traditional planning assumptions and conventional layouts (NHS Confederation, 2005). The recently completed Acad and Becad buildings at Central Middlesex Hospital in London provide one such example, as shown in Figure 3.3. The NHS has evaluated this project in terms of service improvements in their report Local Hospital: Lessons for the NHS, 2008.

Design can also provide efficient and effective layouts that support productive workflows that make best use of staff and technology. Supporting efficient flows

Figure 3.3 The BeCaD (Brent Emergency Care and Diagnostics) at the Central Middlesex Hospital, North West London Hospitals Trust, UK.

for patient and staff journeys, reducing waiting times and making it easy to find places all have implications for design. Many NHS organisations are starting to use Lean principles borrowed from the Toyota experience to rethink their processes and to help get great value from the organisation. The Lean Healthcare Academy (http://www.leanhealthcareacademy.co.uk/) embraces the use of these principles and demonstrates how to achieve service transformation through e-learning, and share information through its newsletter, etc. The implications of Lean thinking for the planning and design of the environment for healthcare are beginning to emerge. Meanwhile, the role of design in the recruitment, retention and perform-ance of NHS nurses was highlighted in a CABE campaign for healthy hospitals, in which 93% of Directors of Nursing stated that the design of the internal environ-ments of the hospital is very important in nurses' performance (Price Waterhouse Cooper, 2004).

Design can help to improve safe working by providing spaces that encourage team work. Design that supports good communications and discourages interruptions can also help to reduce medical errors. There is growing evidence that the number and size of single rooms can affect infection rates, nurse workloads and bed management. Bringing services to the patient rather than moving the patient around is thought to reduce the risk of cross-infection and medical errors. It may also reduce the length of time patients stay in hospital (see National Patient Safety agency research on envi-ronment at http://www.npsa.nhs.uk/nrls/improvingpatientsafety/design/).

The planning of the physical environment can help people to assume respon-sibility for keeping fit, focus on well-being and be more independent. Improving access to health advice in a range of different settings, including supermarkets, community centres and libraries, will be key in helping to integrate health into the wider social infrastructure and to make advice and information more acces-sible. The integration of services has been successfully addressed, for instance, at the St. Peter's Integrated Health and Leisure Centre, located in Burnley, East Lancashire, UK.

Personalisation of services is likely to become more important in future, and environments that value privacy and dignity, and foster a sense of belonging for staff and patients will be vital. Creating a healing environment with good natural daylight, effective use of colour, views, and a sense of place makes a difference to patient satisfaction, recovery times and length of stay in hos-pital, and therefore to overall costs. The question of the value of design has been addressed in several reports, and some of the key factors cited concern regarding reduced use of medication and time spent in hospital, reduced ver-bal outbursts and threatening behaviour in mental health units and improved staff morale (CABE, 2002). CABE (2002) makes the arguments that we cannot afford NOT to invest in good design if we wish to improve the quality of life, equity of opportunity and economic growth. An example of such approach can be seen at the project for the Great Ormond Street Hospital for Children NHS Trust, shown in Figure 3.4.

Figure 3.4 Great Ormond Street Hospital for Children NHS Trust, London, UK.

3.5 Future proofing design

The pace of clinical innovation, plurality of provision and technological advances will result in significant changes to working patterns. It is possible to envisage how models of care will develop and be delivered over a 5-year span. But after that it becomes more difficult to predict. Yet, designing and constructing a building can in itself take many years, with the result that the physical layout could be out of step with clinical requirements even before the new building comes into use (NHS Confederation, 2005).

How can we ensure that what we are building now will be fit for the future? Is it possible to determine the rate of change and how it can be suitably accommodated? It will mean moving away from the idea that models of care are static and that space is a fixed commodity. This implies adopting both a more dynamic planning system approach and a shift in how we think about managing space over time. Scenario planning can help to formulate mental maps of the consequences of different planning approaches. This method can also help to establish how different parts of the building will need to change at different rates and therefore bring a more sophisticated understanding of how to manage and use space more effectively.

It is clear that bespoke solutions that closely fit a specific service model will not offer the greatest flexibility over time. Being able to distinguish what is core business from services that are transient and those that could be provided in other ways is key to unlocking this issue. Linking these different space management needs to the design of the physical fabric is vital. For example, making use of generic rented accommodation for, say, offices may provide a more flexible solution; some services may be managed and run by other organisations in future and the building layout should make this possible; certain services may be best provided from other locations. The focus on buildings that have a more certain life and are core to the delivery of care.

Maximising the potential of the site is key to good master planning. Strategic layout of the infrastructure can minimise restrictions for future developments. So often, designs are hampered by poor site planning, which offers unnecessary constraints and sterilises the sensible site development now and in the future. It is rare for proposals to consider what will happen to the site when the new proposed buildings come to the end of their useful life.

Different parts of the building structure and fabric have different lifespans, and these can be designed to be able to replaced or changed at different times; so whilst the structure may have a 30–60-year lifespan, engineering services may have only a 10–15-year lifespan, and internal spaces may need to change every 2–5 years. Sometimes, it is possible to design and build 'shell space' that can be fitted out over time as the specific requirements become clearer.

Grouping functions with similar engineering requirements can offer more economic layouts, together with being more adaptable to change. Structural grids, engineering services and ceiling heights can be designed to suit particular activities rather than having to be the same throughout the building. One size does not necessarily suit all. Highly serviced environments will have more technical requirements and be more expensive to build than, for example, consulting rooms or offices. Different clinical activities are likely to change at different rates and need to be able to change more frequently than, say, consultation. They need to be able to change without impacting detrimentally on adjacent spaces and services. Creating buffers of 'soft space' between the hard-wired, highly serviced spaces can provide important elbow room for expansion and change. Deep plan spaces do not necessarily offer the most flexible plans.

There are currently thousands of different room types for health buildings, and it is estimated that this could be significantly reduced. Clearly there are advantages to building fewer room types that can be used interchangeably over time as service requirements change. Managing space as a resource rather than as territory will improve space utilisation rates and will almost certainly require changes in working practices to become fully adopted.

One area that is often neglected in health care planning is the public space such as corridors and spaces in between the clinical functions. The layout of the vertical and horizontal circulation is significant in the planning for future expansion and flexibility of use. These spaces usually contain elements such as staircases and lifts that are expensive and disruptive to move or replace. The positioning of these can not only affect the productive working processes but also allow the building to retract or expand and even allow different parts of the building to come under different tenures over time. Developing an understanding about the hierarchy of circulation routes is key to providing suitable connections between departments and spaces that can allow the combined or separate use by patients, staff and goods.

3.6 Design matters

There is growing evidence showing the impact of the design of the healthcare environments to staff performance, improvements in patient health outcomes and staff and patient safety. Further work is now focusing on the potential for the wider built environment of our neighbourhoods and cities to impact on healthy lifestyles. This means, for example, creating attractive places that encourage active travel such as cycling, walking and exercise, in general. It implies more strategic planning of towns to create networks of community facilities close to home in neighbourhoods that are planned, for example, with less reliance on care travel.

The NHS has developed a Design Review programme for major capital projects to ensure that good design is embedded in the NHS building programme. Managed by the Department of Health Design and Costing team and supported by CABE, the panel reviews design proposals at three key stages in development. A panel consisting of architects, engineers and planners meet the trust to review the proposed scheme designs and make constructive criticisms and recommendations.

Until now this programme has focused largely on the major hospital projects. But with the shift towards more community-based services, it is anticipated that it will now focus on more community-based projects such as community hospitals, more mental health facilities and the larger projects for primary care facilities.

Key themes arising from the reviews, endorsed by findings from an independent evaluation of the review process and a recent refresh of the criteria, have been identified as follows:

- Master planning: the relationship of the health building with its context.
- Quality of design: in particular the design of the internal environment to provide a healing environment and positive experience for users.
- Sustainable development: the agility of the design to be flexible to change and ability to respond to the needs of climate change.

3.7 Measuring design quality

A systematic appraisal framework for healthcare design has been developed with the aid of a toolkit called AEDET evolution. It is based on the 'Vitruvian' principles of 'commodity, firmness and delight' that have been translated into modern parlance as functionality, building quality and impact. The NHS toolkit closely follows the Design Quality Indicators developed by the Construction Industry Council as an industry standard for evaluating design (NHS Confederation, 2004).

AEDET provides a useful tool for client and their advisors to evaluate and rate the degree of design excellence. It is a reminder to those using it that the environment is a complex organism with no absolute right answers. It has also put issues such as ambience, light quality and sensory stimulation on an equal footing with 'fitness for purpose' (function) and technical performance (engineering and construction). AEDET has become a mandatory element for many procurement processes. There is a move for it to be developed as an independent objective assessment, with the requirement for the achievement of an excellent rating as part of the business case approval process.

The NEAT environmental toolkit focused on the environmental impact of design. This has typically encompassed energy and water conservation, bio-diversity and waste disposal. Widening this to include not only the physical attributes of the design proposal but also its social impact is now underway. This has been developed as a 'BREEAM for Health' tool managed by the BRE (Building Research Establishment). Once again, an excellent rating will be a minimum requirement for new buildings for business case approval (see BREEAM for Health, http://www.bre.co.uk).

Each NHS Trust is expected to appoint a design champion at board level, someone whose role is to raise awareness about design and support project developments. This expertise needs to be balanced with sufficient expertise in the project teams to be really effective.

Commission for Architecture and the Built Environment (CABE) and the RIBA (Royal Institute of British Architects) both offer support services to clients about design. CABE has an 'enabling' programme in which up to 10 days of expert design input can be assigned to a project. The RIBA Client Design Advisory service also has a network of experts who can be commissioned to support the client through design development.

It has become apparent from these and other initiatives that the more able the client is to articulate and communicate the design brief, the better the quality of the resulting scheme. There is a call for greater emphasis on design at the start of the programme, and the process of 'design exemplars' is now being explored in relation to both partnership and more traditional procurement methods. There is no doubt that well-designed buildings rely on a combination of informed clients and imaginative architects, as it is has been the case in the design of the Bamburgh Clinic (Figure 3.5).

Figure 3.5 Bamburgh Clinic St Nicholas Hospital, Newcastle, UK.

3.8 Final remarks: making places

Good-quality design has enduring value and it is as important in making the space effective and flexible. Smarter strategic investment that can help to support the reform of care delivery, where more productive and safe patient care is needed, is essential.

Being good custodians of the environment through design is essential to sustaining prudent use of the environment through energy and water conservation, selection of materials, bio-diversity and waste disposal. Master planning ensures the strategic use of land, makes best use of the site and offers the potential for health to be more integrated into t he civic realm. Making places that lift the spirit, provide greater user control, and help to develop a sense of belonging are important for all users – whether they are patients, visitors or staff.

The healthcare sector is undergoing significant and rapid changes. The policy context, drive for more effective and productive delivery of services, is happening alongside investment in buildings that are fit for the future and deliver optimistic and quality settings for care. The need to plan for uncertainty and design for change is imperative. Perhaps we can do better than strive for 'long life, loose fit and low energy'[2] buildings.

References

CABE (2002) *The Value of Good Design*, Commission for Architecture and the Built Environment (CABE), London.

Cavill, E. N. (2007) *Building Health: Creating and Enhancing Places for Healthy, Active Lives: What Needs to be Done?*, National Heart Forum, in partnership with Living Streets and CABE, London WC1H 9LG.

DoH (2000) *The NHS Plan: A Plan for Investment, a Plan for Reform*, Department of Health, National Health Service, The Stationary Office, London. Available at: http://www.dh.gov. uk/en/Publicationsandstatistics/Publications/PublicationsPolicyAndGuidance/DH_4002960

DoH (2005a) *Creating a Patient-Led NHS: Delivering the NHS Improvement Plan*, Department of Health, London. Available at: http://www.dh.gov.uk/en/publicationsandstatistics/publications/publicationspolicyandguidance/dh_4106506

DoH (2005b) *Commissioning a Patient-Led NHS*, Department of Health, London. Available at: http://www.dh.gov.uk/en/Publicationsandstatistics/Publications/PublicationsPolicyAndGuidance/DH_4116716

DoH (2006) *Our Health, Our Care, Our Say: A New Direction for Community Services*, Department of Health, London. Available at: http://www.dh.gov.uk/en/Publicationsandstatistics/Publications/PublicationsPolicyandGuidance/DH_4127453

DoH (2008) *High Quality Care for All: NHS Next Stage Review Final Report* (Lord Darzi Report),The Stationery Office, London. Available at: http://www.sor.org/news/files/images/Darzi_Final_Report.pdf

Foresight (2008) *Mental Capital and Wellbeing Project Outputs*. Available at: http://www.foresight.gov.uk/OurWork/ActiveProjects/Mental%20Capital/Welcome.asp

Francis, S. and Glanville, R. (2001) *Building a 2020 Vision: Future Healthcare Environments*, The Stationery Office, Nuttfiled Trust, London.

Lawson, B. and Phiri, M. (2003) *The Architectural Healthcare Environment and Its Effects on Patient Health Outcomes: A Report on an NHS Estates Funded Research Project*, The Stationery Office, London.

NHS Confederation (2004) *FHN Briefing Ensuring Good Design for Healthcare*, The NHS Confederation, London SWIE 5DD.

NHS Confederation (2005) *FHN – Futures Healthcare Network – Briefing Optimising Design*, The NHS Confederation, London SWIE 5DD.

NHS Confederation (2006) *FHN – Futures Healthcare Network – Briefing Sustainable Communities*, NHS Confederation, London SWIE 5DD.

Price Waterhouse Cooper (2004) *The Role of Hospital Design in the Recruitment, Retention and Performance of NHS Nurses in England*, CABE, London. Available at: http://www.cabe.org.uk

Ulrich, R. S., Zimring, C., Zhu, X., DuBose, J., Seo, H., Choi, Y., Quan, X. and Joseph, A. (2008). *A Review of the Literature on Healthcare Design*. Available at: http://www.healthdesign.org/hcleader/HCLeader_5_LitReviewWP.pdf, Last accessed Sept 2008.

---

[2]Phrase devised by Alex Gordon, former President of the RIBA 1964.

Designed with Care?
The Role of Design
in Creating Excellent
Community Healthcare
Buildings

Kate Trant

4.1 Introduction

The issue of designing buildings for healthcare has been discussed in Chapter 3, and this chapter provides some further thoughts on the area, and more specifically in the importance of healthy communities. In the context in which those responsible for commissioning, designing, constructing and managing health facilities now operate, it is essential that each building is considered within the context of creating places, spaces, neighbourhoods for better health and *well-being*. Only then can the next significant strides in creating and supporting healthy communities be made. Healthcare buildings have particular requirements that make them worthy of focused study, but never in isolation from their context and their contribution to creating and supporting healthy communities.

> *The design and management of the built environment plays a pivotal role in promoting and sustaining health and, in particular, in tackling rising levels of inactivity.*
> (CABE, 2006a, p. 2)

In the UK currently, we are witnessing three major health and well-being concerns:

1. obesity;[1]
2. the ageing population – the demands placed on health and social care by the UK's changing demographic. The number of those in the UK suffering from some form of dementia has increased to a current figure of around 700 000;[2]
3. mental health – the increasing recognition of the benefits of creating places and spaces that enable improvements in our mental well-being (see Halpern, 1995; Foresight, 2008).

These three concerns sit under the overriding issue of environmental sustainability and how to create thriving, healthy communities that have a positive environmental impact. Work amongst those agencies[3] involved in each of these health and well-being areas, as well as those involved in creating and managing new and existing places and spaces, continue to work towards establishing the beneficial links between the built environment and each of these broad issues (Jackson, 2003).

This chapter looks at the role of design in creating individual buildings, as well as how the design of high-quality spaces and places contributes to creating healthy neighbourhoods. It will consider how successful healthcare building design can support those providing care and create an environment in which access to healthcare services is made easier. The consideration of access to healthcare extends to looking at how the wider built environment can enable a healthier lifestyle and, therefore, improve health. Finally, these themes will be illustrated by looking at a selection of neighbourhood healthcare building cases.

4.2 Why does design matter?

Good design is more than aesthetics or style. It is about ensuring that a product – a piece of equipment, a building, a neighbourhood, a city or even a service – makes life better for anyone who comes into contact with it. Building for healthcare involves design as a holistic exercise that combines a wide range of ingredients that, working

---

[1] For an overview of the current trend towards obesity in the UK, *see Foresight Tackling Obesities: Future Choices – Modelling Future Trends in Obesity & Their Impact on Health*, 2nd ed., Government Office for Science, 2007. Available at: http://www.foresight.gov.uk

[2] See the pioneering work done by the Helen Hamlyn Centre at the Royal College of Art, London, on the design implications of an ageing population, http://www.hhc.rca.ac.uk

[3] UK agencies engaged in looking at the potential for the built environment and design to impact positively on health and well-being include CABE (Commission for Architecture and the Built Environment), Foresight, HUDU (Healthy Urban Development Unit), Living Streets, National Heart Forum, NICE (National Institute for Health and Clinical Excellence), UKPHA (UK Public Health Association), WHO Collaborating Centre for Healthy Cities and Urban Policy, and University of the West of England, Bristol.

well together, make a healthy building. All neighbourhood healthcare buildings should provide high-quality facilities for patients, their families and carers, and staff. Organisation of space, careful consideration of adjacencies, and design for ease of maintenance and cleaning are all aspects of designing for healthcare that go beyond simple aesthetics.

A well-designed healthcare building should be functional, well made and beautiful. We must also consider whether a building makes a positive contribution to the community in which it is based. Doctors' surgeries, clinics, medical centres and healthcare buildings of all types are community buildings; they serve local residents and should make a positive contribution to their surroundings. The successful healthcare facility is a cornerstone of community health.

4.2.1 Building healthy neighbourhoods

Although there is an increasing understanding of relationship between the built environment and health and well-being, using the built environment for the promotion of health is still too often seen as merely a question of where to locate health facilities, rather than how to use the built environment as a tool for reducing demand on those facilities. With a renewed focus on *prevention* comes an increased emphasis on creating environments that encourage healthy lifestyles, for example, greater levels of walking and cycling – using the built and natural environment to improve our quality of life as well as our health. This means improving the design and management of all aspects of our built and natural environment, including our streets and neighbourhoods, transport infrastructure and workplaces, parks and outdoor spaces (CABE, 2006a).

> . . . the way we build our cities, design the urban environment, and provide access to the natural environment can be a great encouragement – or a great barrier – to physical activity and active living. That is what we need – encouragement. We need to make it easier to be physically active in our everyday life at work, at home, at school, in our neighbourhoods . . . (National Heart Forum, 2007, p. 5)

The quality of the design of the built environment as a whole and its impact on our health has to be seen as part of the same endeavour as designing high-quality healthcare buildings with a positive impact on health outcomes, rather than as two separate exercises. With Britain having the highest number of obese people in Europe (DoH, 2007), thinking about how we can improve our general health and fitness is as important as thinking about how we are treated when we are sick. Most sustained exercise is taken just doing everyday activities, such as travelling to work or going to the shops, rather than specifically to get fit (CABE, 2006a). If towns and cities feel and look good, more people will choose to walk and cycle. So, those commissioning, designing and constructing new homes, education facilities, public spaces, healthcare buildings and so on around the country, as well as managing existing places and spaces, have a responsibility to make healthy behaviour an easy and attractive option.

4.2.2 Access to health

Design is still regarded by some as an area too subjective for measured research and judgement. It is now, however, widely accepted that the impact of our immediate environments on our productivity levels, our capacity to relax, our ability to easily navigate where we are and on giving us opportunities to effectively engage with one another is a crucial factor in the success or otherwise of the places where we work, live and play (CABE, 2005a). In the area of healthcare, activities such as these benefit from taking place in healing environments that not only are fit for purpose but also better enable the improved delivery of a variety of services, by allowing for efficiency, flexibility of use and simple control of comfort levels.

The expression 'healing environment' is generally used in the UK to describe healthcare buildings that *underline* rather than *undermine* the care that is given in them, supporting clinical activities by providing humane places and spaces that sooth, calm and reduce anxiety in those receiving care. They also invigorate those receiving treatment, and will support those whose job it is to provide care.

Patients deserve to receive care within healing environments that are uplifting whilst satisfying clinical requirements, and staff should be confident that they are being helped in providing the best care. Nonclinical services have similar requirements, and bringing together these activities within thoughtfully designed premises is beneficial to all concerned. Meanwhile, a building's appearance is important. A health centre that is neither physically nor culturally accessible, with an exterior that does not address its context and is not appropriate to the community it serves, will not make a positive contribution to its locality. A surgery that is hard to maintain and quickly looks shabby, that might function satisfactorily but, for example, is lost in a sea of cars (with everything that this implies about the local transport infrastructure), will do nothing to lift either its surroundings or the spirits of its users.

While our healthcare buildings need to demonstrate to the communities that use them that their health and well-being is important, they also need to communicate to the people who work in them that their role is valued. A study amongst nursing staff demonstrates that the quality of design has an impact on recruitment, retention and performance (CABE, 2004):

- 78% of Directors of Nursing say that hospital design impacts on the recruitment of nursing staff, with external space and internal environment the most influential design factors on recruiting nurses.
- 87% of nurses say that working in a well-designed hospital would help them do their job better: 'You get an impression when you walk in . . . if it looks scruffy this may reflect the management style'.

Any healthcare building has multiple functions. Success depends on satisfying a complex set of requirements; design has a crucial role in delivering a building that responds well to the needs of all its users.

4.2.3 Surprise and delight

As the UK's recent ongoing healthcare building programme has unfolded, what is evident is a general ability to deliver on most functional requirements; what is lacking, or at least inconsistent, is an ability to translate into the built form our increasing understanding that the quality of our health buildings impacts on the care that is delivered. The role of a healthcare building in lifting the spirits as well as to be functionally successful, though increasingly understood, has still to become a real basis for its design.

These more intangible characteristics of a building, whilst difficult to measure and demonstrate, are essential in supporting and complementing the clinical care that goes on inside them. A recent study demonstrated that the public appreciate the importance of high-quality design (CABE, 2002):

- 85% of people agree that better quality buildings and public spaces improve the quality of people's lives.
- 85% of people agree that the quality of the built environment makes a difference to the way that they feel.

4.2.4 Designed with care

The following cases illustrate the importance of high-quality design of healthcare buildings, as well as their contribution to their immediate neighbourhood. They can all be found in CABE's survey of neighbourhood healthcare buildings, *Designed with Care* (CABE, 2006b).

Maggie's Highlands in Inverness (Figures 4.1 and 4.2), designed by Page\Park from Glasgow, clearly demonstrates how the use of form, materials and colour contribute to creating a building that communicates 'care', as well as provides a service. The design of the building contributes to the warm, welcoming feel of the centre, and builds on the belief that health centres are as much for information and support as for cure.

The first centre opened in Edinburgh, a year after the death of Maggie Keswick Jencks. Maggie did much of the initial planning for the Edinburgh centre, saying, 'We want to make spaces which make people feel better, rather than worse' (CABE, 2006b).

4.2.5 Open all hours

With the shift in the healthcare agenda to a focus on health and well-being, the community function of our healthcare buildings has increased in importance and is under close scrutiny.

One of the newer models of delivery of care is the walk-in centre. Luton Walk-in Centre holds a highly visible town centre location and key to the success of this building is its accessibility. Its design, by David Morley Architects, reinforces that this building is there for the local community. The first person was queuing at the door when the building opened in February 2004, and demand for its services remains high.

Figure 4.1 Interior at Maggie's Cancer Caring Centre, Inverness, designed by Page\Park. First-time visitors and regulars meet here to chat informally or to talk to members of Maggie's staff. Photograph taken by A&M Photography.

Figure 4.2 The sitting room at Maggie's Cancer Caring Centre, Inverness, designed by Page\ Park. Photograph taken by A&M Photography.

4.2.6 Better isn't good enough

A commonly heard refrain is that a new healthcare building is 'better than it was before'. This is particularly evident currently, where we are seeing many instances of combinations of previously separate services brought together in a new building from smaller, poor quality locations. On initial occupation, staff and patient satisfaction is understandably high. Nevertheless, however well received a new building might be on completion, the degree of its success will change over time, and ongoing evaluation of its performance is required for any real assessment.[4] This is particularly pertinent where a building has a role in a changing, rather than replicating, existing culture.

Hove Polyclinic by Nightingale Associates (Figures 4.3 and 4.4) demonstrates how high-quality design can be instrumental in bringing together successfully a

Figure 4.3 Glass blocks light at the curving staircase at Hove Polyclinic by Nightingale Associates. Photograph taken by A&M Photography.

range of different health services under one roof that have not previously been offered in a single building. A long process was involved in this project, involving the practice working closely with the two healthcare trusts who were clients for the building (South Downs and Brighton), and clearly shows how a good, brief consultation and collaboration are crucial to success.

4.2.7 Must try harder

Few clients set out to design a bad building that does not contribute positively to its users or its surrounding communities. However, in the UK, healthcare buildings are so often characterised by mediocrity. This unintended consequence results from a range of factors, including the following: a lack of understanding of good design and the value that it brings to a project; speed of project delivery that does not allow

Figure 4.4 The atrium at Hove Polyclinic, by Nightingale Associates, brings light into the first floor waiting area. Photograph taken by A&M Photography.

---

[4]CABE is promoting postoccupancy evaluation for healthcare buildings; working with the Department of Health to develop systems for assessing value for money in LIFT projects (Local Finance Improvement Trust); and encouraging pre-project evaluation for existing premises, using a version of the Design Quality Indicator tool (see http://www.cabe.org.uk).

for sensible and measured consideration of how design can deliver on a range of requirements; or little understanding of how service planning will result in a building.

A healthcare building is, by nature, a community building and its value over time is complex. At the very least, it has social and environmental – as well as economic – value. Some of the benefits of a well-designed building can be measured. However, it is typically more difficult to measure social and environmental benefits and to retain these as fundamental objectives. In a cultural climate in which short-termism is often dominant, it is important to remember that the contribution of a health building locally is based on criteria far more complex than its immediate economic value alone.

4.3 What makes a good healthcare building?

CABE has developed a set of key elements for good healthcare buildings (CABE, 2005b, 2006b). These are described in the following subsections.

4.3.1 Good integrated design

Design excellence is not just about attractive buildings; good design also takes into account how a building contributes to its environment (Figure 4.5). For example,

Figure 4.5 Advance Dental Clinic, Chelmsford, by Richard Mitzman Architects. The design of Advance Dental Clinic by Richard Mitzman Architects makes the best use of its small site in a residential suburb of Chelmsford. Photograph taken by A&M Photography.

wherever possible, this should include ease of access and straightforward integration with public transport (Figure 4.6).

4.3.2 Public open space

There is an increasing body of evidence that tells us that nature can contribute to healing. A healthcare building should extend its concern for patients, staff and visitors by providing well-managed public space. Pulross Intermediate Care Centre offers accessible gardens in an inner-city environment for use by patients and their visitors, and staff.

4.3.3 A clear accessible plan with one main reception

Clear wayfinding based on a logical plan can contribute positively to the experience of a patient or visitor, starting with a clearly placed entrance, with easy access for all (Figure 4.7). Signage is too often used to compensate for a badly designed building that does not offer intuitive wayfinding (Figure 4.8). Reception areas and information points are key to orientation; resolving the conflicting requirements of the reception function of any healthcare building is a common challenge for healthcare design.

Figure 4.6 Roadside approach to Idle Medical Centre in Bradford by VJQ Architects. Photograph taken by A&M Photography.

Figure 4.7   The gull-wing roof of City Road Surgery, Hulme, Manchester, by Stephen Hodder Architects gives the surgery its distinct appearance. Photograph taken by A&M Photography.

Figure 4.8   Main staircase at Bart's Breast Care Centre, London, by Greenhill Jenner Architects, with David Batchelor's 'West Wing Spectrum' installation. Photograph taken by Nigel Greenhill.

4.3.4 An environmentally sensitive approach to building design, materials, construction and management

Taking advantage of the natural environment can help create a sustainable building, and current technologies and knowledge should be used to maximise a building's environmental credentials. For example, orientation can be used to the best advantage. At Rutland Lodge, the building faces southwest, making most use of the winter sun, but is shaded by trees in the summer (Figure 4.9). Grassroots (Figure 4.10) uses solar panels that provide 30% of the building's power at full efficiency, while rainwater is used to irrigate the grass roof and provide grey water. The energy use can be watched from inside the building on a set of display panels.

Figure 4.9   The waiting area at Rutland Lodge in Leeds by OSA Architects. Photograph taken by A&M Photography.

Figure 4.10 Grassroots, London, by Eger Architects, where the majority of the building is set beneath a grass roof. Photograph taken by A&M Photography.

4.3.5 Circulation and waiting areas

Well-planned waiting areas can help to relax patients, while views out of the building can aid relaxation (Figure 4.11). The choice of furniture can contribute to making the experience a welcoming one, and support the design quality ethos of a well-designed healthcare building.

4.3.6 Materials, finishes and furnishings

Materials and finishes need to be robust and easy to maintain, as well as attractive. Well-selected, fit-for-purpose furnishings will complement a clear approach to design. Conversely, badly chosen materials, fixtures and furnishings will very quickly undermine the work of any architect or client in championing design quality.

4.3.7 Natural light and ventilation

Good-quality light and ventilation are often cited as important by patients, staff and visitors. Both qualities contribute to a calm, comfortable and relaxing environment, as well as contributing to energy efficient building use (Figures 4.12 and 4.13).

Figure 4.11 The view across the Aire Valley from the upper waiting area at Idle Medical Centre, Bradford, by VJQ Architects. Photograph taken by A&M Photography.

4.3.8 Storage

Good provision of storage contributes hugely to the success of a healthcare building. A well-maintained, clean and tidy building communicates a positive message to its users. Good storage contributes to this – 'a place for everything and everything in its place'.

Advance Dental Clinic maximises the use of well-planned storage. Dentist Andrew Moore understands how providing a service in a high-quality environment improves the experience for his patients. The clinic has two surgeries so that one can be cleaned while Andrew is treating a patient in the other. The architect,

Figure 4.12 The use of roof lights at Advance Dental Clinic, Chelmsford, by Richard Mitzman Architects ensures that the building is full of natural light. Photograph taken by A&M Photography.

Richard Mitzman, has calculated that the extra hour gained each day by doing this gains about six weeks every year, quickly producing a return on the dentist's investment.

4.3.9 Adapting to future changes

Space should be viewed as a resource, not a territory, allowing patterns of use to evolve over time. The rapid demand for change and for fresh approaches to the delivery of healthcare at the neighbourhood level means that buildings today need the capacity to adapt to future change, with built in flexibility.

Figure 4.13 View from the upper balcony at Idle Medical Centre, Bradford, by VJQ Architects showing the planted atrium. Photograph taken by A&M Photography.

4.3.10 Out-of-hours community use

Similarly, changing needs of local communities mean that healthcare buildings need to be designed with the possibility of use 24/7 in mind. At Grassroots, in Memorial Park in Newham, a process of public consultation resulted in genuine community involvement in the development of the buildings. By the time the designs for Grassroots were being developed, it was local residents involved in the consultation who were demanding a high-quality landmark building. With the building complete and in use, the same local residents and groups use the building for different community activities (Figures 4.14 and 4.15).

Figure 4.14   Facilities at Grassroots in London by Eger Architects include a versatile multipurpose hall. Photograph taken by A&M Photography.

Figure 4.15   The Healthy Eating Café at Grassroots in London by Eger Architects. Photograph taken by A&M Photography.

4.4 Final remarks

While the focus of this chapter is the design of healthcare facilities, it is essential to consider these buildings in terms of their relationship to the communities they serve, how they contribute to care locally and their role in improving health and well-being. Understanding the link between the built environment and health is nothing new. Many of our homes, cities and suburbs were designed to combat health problems. But increasingly, the potential offered to health and well-being by a high-quality built environment is understood. There is a growing recognition that people get better quicker in well-designed healthcare buildings, but now that the public health agenda in the UK is moving away from treating illness to preventing it, the opportunities are far greater than looking solely at the design of buildings.

The health of all of us in the UK depends not only on the care we receive from our healthcare professionals, but also on the environments in which we live, work and play (CABE, 2006b).

References

CABE (2002) *Streets of Shame Executive Summary,* CABE – Commission for Architecture and the Built Environment, London.

CABE (2004) *The Role of Hospital Design in the Recruitment, Retention and Performance of NHS Nurses in England*, CABE, London, UK.

CABE (2005a) *The Impact of Office Design on Business Performance*, CABE and the British Council for Offices, London, UK.

CABE (2005b) *Lewisham Primary Care Trust Children and Young People's Centre: Design and Innovation for Primary Health and Social Care*, CABE, London, UK, p. 7. Available at: http://www.cabe.org.uk/default.aspx?contentitemid=665

CABE (2006a) *Physical Activity and the Built Environment*, CABE, London, UK, p. 2.

CABE (2006b). *Designed with Care: Design and Neighbourhood Healthcare Buildings*, CABE, London.

DoH (2007) *Health Profile of England*, Department of Health, XX.

Foresight (2008) *Foresight Mental Capital and Wellbeing Project,* The Government Office for Science, London. Available at: http://www.foresight.gov.uk/OurWork/ActiveProjects/Mental%20Capital/Welcome.asp

Halpern, D. (1995) *Mental Health and the Built Environment: More Than Bricks and Mortar?*, Taylor & Francis, London.

Jackson, R.J. (2003) *The Impact of the Built Environment on Health: An Emerging Field. American Journal of Public Health*, Vol. 93, No 9, 1382–1384.

National Heart Forum (2007) *Building Health: Creating and Enhancing Places for Healthy, Active Lives. What Needs to be Done?*, National Heart Forum, London WCIH 9LG.

The Stages of LIFT – Local Improvement Finance Trust and Delivery of Primary Healthcare Facilities

Richard L. Groome

Public-sector procurement and construction have gone through various evolutionary changes in recent years. The old method of Joint Contracts Tribunal (JCT) conventional build contracts in the UK, such as JCT 98 and earlier, had carefully defined roles for all the participants, for example the architect who would have, depending on the version of JCT being used, a duty to provide all design information to the contractor including, at one end of the spectrum, how to actually build the project and at the other, what height to put the light switches. The problems with these contracts were that conflicts arised if the duties prescribed were not fulfilled by any one person. Risk was not shared; it was allocated, and again the slightest misinterpretations would lead to problems. The risk of cost overruns and the funding thereof was much more with the public sector than it is today, and the responsibility of getting the initial specification correct was squarely with the public sector. The overall judgement on this period of procurement would be that it was too long, too adversarial, too restrictive and too risky for the public sector.

Two new approaches have changed this: firstly, the move to privatising the finance of major projects so that the funding is now mainly private sector and, secondly, the adoption of a process approach. The latter appears quite complex but in fact maps the whole project and reduces the occurrence of conflict situations. For public procurement this means the adoption of design and construct (D&C) contracts, which respond to an output specification rather than an input specification, and cost risk is transferred largely to the private sector. The private sector can also take service delivery risk, for example cleaning, maintenance, and waste removal, but the final delivery of the core service (e.g. healthcare delivery) remains with the public sector.

These contracts have typically been placed under the Private Finance Initiatives (PFI), Procure 21 or Local Improvement Finance Trusts (LIFT), which are guaranteed maximum price contracts (see Section 5.2.6). Their structure, however, derives from the work done by British Airports Authority to use the New Engineering Contract, particularly at Heathrow Terminal 5, and earlier work on zero defects championed by Sir John Egan at the same company. This contract, with its shared risk pot, encourages innovation and shares risks and gain, whereas the PFI and LIFT contracts may not, but the basics are the same. There are barriers to innovations related to such contracts in the value-for-money area, where an over-emphasis on short-term capital costs often rules out long-term savings. However, in the health sector, organisations such as Community Health Partnerships have championed new, improved contracts to strengthen such weaknesses.

In this chapter the complex process to take a LIFT private–public partnership (PPP) project through to completion and operation will be described. Although tortuous, it offers huge advantages for public procurement in that all future maintenance and replacement of assets are included in the project agreement, releasing the public client to concentrate on the core service delivery. Such description is based on the authors' practical experience in delivering UK LIFT schemes, from a private sector's perspective.

5.2 The LIFT process

This section describes a typical LIFT procurement in the health sector. LIFT represents a true joint venture, public and private sector, for the life of projects up to 30 years. LIFT is used typically for projects under £30m, and involves the formation of a joint venture company with an equity structure 40% public–60% private for any dividends or future rewards (a PPP). A LIFT Company (LIFTCo) is formed to act as the delivering vehicle for new facilities.

There is nothing new with the involvement of the private sector in the delivery of public services. PPPs in the form of build, operate and transfer were used as early as 1858 for the construction of the Suez Canal (Cartlidge, 2006). The capital funding of new projects is usually 90% private, however. The design stage approval process for LIFT schemes is graphically presented in Figure 5.1 and described in the following subsections.

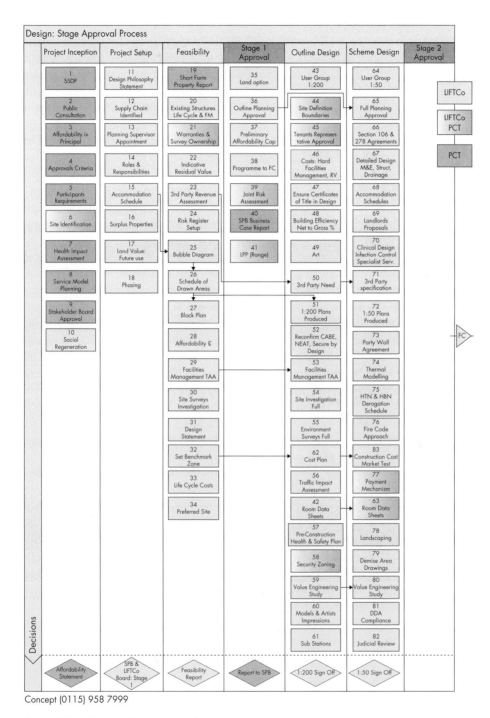

Concept (0115) 958 7999

Figure 5.1   Design–stage approval process.

5.2.1 Project inception

In the health sector, the local health guardians such as primary care trusts, strategic health authorities and local councils start the process by agreeing to a strategic service delivery plan (SSDP). This document concentrates on replacing or renovating the current real estate and providing specific services, in particular areas, to improve the health of the population. In Manchester, for example, the SSDP in 2004 wanted to address, inter alia, the lowest life expectancy for men in England, poor immunisation rates for children, very poor disabled access to existing premises, most buildings being undersized and many unfit for purpose.

The SSDP leads to the ability to prioritise projects, and is the first stage in identifying the projects that will go through the procurement process. In order to get a full picture, a primary care strategy including service models should be in place, and a health impact assessment for the project is advised. The health impact assessment describes what health benefits will be realised if the project proceeds. Both of these may involve public consultation. A useful assistance to the project will be a need to regenerate an area or society, hence the link to local councils and regeneration companies. Particularly in LIFT, the initiative was conceived to give a real 'lift' to local communities. At the end of this process, a *site* should be roughly earmarked.

Next the commissioner(s) need to state their outline requirements and approvals criteria. The commissioners might be the primary care trusts, local councils, community groups or in some cases commercial organisations like pharmacists. They also have to indicate their affordability constraints – on the basis of the services they need to provide, how much can they pay for the property/facility?

Lastly, the local LIFTCo will give an initial response on all of the above in terms of size of building, complexity and cost, and all parties should then ask for an outline approval to proceed from their stakeholders or boards. At this stage, under the LIFT rules, the costs to the LIFTCo are at its risk.

5.2.2 Project set-up

The next stage is mainly focused around the definition of roles, responsibility and risk sharing. A project risk register will list what risks will be taken by the public sector; what lies with the LIFTCo and what it intends to pass to its supply chain. The supply chain will be identified at this stage – often the main suppliers will be in framework agreements[1] if they have already completed work for the partnership. These include cost improvement and value for money targets typically; if not, a future tender may be required.

The output specification work will have started with a schedule of accommodation provided by the commissioners, and a tenant's requirements document with

---

[1]Framework agreements are long-term relationships with suppliers. When procuring over a period of time, a framework can deliver benefits, such as reduced transaction costs and continuous improvement within long-term relationships, better value and greater community wealth solutions (see http://www.constructingexcellence.org.uk/tools/frameworkingtoolkit/whyuse.jsp for details).

design philosophies. Please note that under the latest LIFT documentation, much of this needs to be pre-designated by the UK Department of Health.

With the target site or area in mind, both the LIFTCo and its customers (PCTs, local authorities etc.) will look at surplus properties in the area (for use, demolition or disposal) and the residual values of properties and land, to see if a favourite site presents itself. A new site may need to be sourced and acquired. Then the key persons in the project will be nominated form all organisations involved, responsibilities and authorities clearly stated and communications protocol agreed. One often forgotten and key role is what used to be called the planning supervisor under the Construction Design and Management (CDM) Regulations. This is now termed the CDM coordinator, and it is a client responsibility and a legal requirement that such a coordinator is assigned before any design is started. This person monitors the construction and design process independently, with a particular emphasis on health and safety.

5.2.3 Feasibility

In the final stage before the first formal approval, further work is done to gather as much information as possible without incurring too much cost for either public or private sector, in case the project is find not to be viable in the future. Examples of this are follows:

- Reports will be generated on existing properties with future life cycle and FM costs estimated and residual values determined.
- Early massing plans (block areas) for the proposed development will be developed for checking the overall size of individual departments and the need for any adjacencies.
- Site surveys of likely sites will be carried out.
- Benchmark costs from similar constructions will be tabled.
- The future requirements of FM, life cycle and site services will be established.
- Any third-party revenue coming into the project (like from retail outlets within the building) will be estimated.
- Any restrictions or covenants on possible sites will be explored.
- The cost and affordability will be made as firm as possible.

At the end of this process a preferred site is mandatory, and an outline planning application needs to be started.

5.2.4 Stage 1 approval

This is very similar to what UK local authorities call Gateway 2, under the rules of the Office of Government and Commerce. This is a formal approval by the local strategic health authority and/or the local council to proceed to the detailed design stage and undertake costs, given to the LIFTCo. To obtain it, the LIFTCo must fulfil the following:

- be in control of the land (which suggests at least an outline planning permission be in place);
- have cognisance of all the user requirements and guarantee to meet them;
- have offered a 'lease plus payment'[2] that is *affordable*.

The PCT or council will have similarly produced a business case to match the above, at the same time, with risks assessed and a programme for the project laid down.

Once the above conditions are met for a project identified as in the SSDP, then stage 1 approval is normally given. It is interesting to note that as yet, no detailed design has been developed. Nonetheless, the successful LIFT projects are those that follow this process carefully: Time and cost should be saved at later stages, and the future service delivery is more likely to be what was intended.

5.2.5 Outline design

At this stage the more detailed design, and incurring of costs, commences. Firstly, the site boundaries and title have to be confirmed. Property matters are the bane of LIFT projects, and the LIFTCo legal team should now be involved to produce a Heads of Terms document, involving what is intended, on what property and what deals and documents will be involved.

User group meetings will be convened, or in the case of some projects, re-convened from earlier work. By a combination of the input from these, and specialists such as healthcare planners and architects, 1:200 drawings of the proposed site are produced. We found it useful on the Manchester projects to project these drawings onto screens, and to have computer-aided design (CAD) specialists working alongside as part of the sessions, so that suggested changes could be made immediately.

Site investigations and surveys, particularly environmental, are now required in detail; and the LIFTCo design team will commission the BREEAM (Building Research Establishment Environmental Assessment Method) survey, which is now part of UK Building Regulations Approval. Normally, a rating of BREEAM 'Excellent' will be the minimum expected, and this will involve assessing all the environmental impacts of the new development. There will also be a need to add any input from 'Secure by Design' (designing for better security).

The facilities maintenance (FM) assumptions are now worked through; usually these are called technical asset assumptions (see later) and are a statement of what the technical plant in the new health centre has to achieve. These will

---

[2] The lease plus payment is the monthly charge made by the LIFTCo to its tenants, which includes the lease of the premises and all maintenance/life cycle charges. In LIFT, there are no dilapidation clauses as in commercial leases; the cost of renewal of the asset is included in the charge. Also, the LPP can only increase by retail price index (RPI) inflation, so commercial rent reviews do not apply. All these factors give cost certainty to the tenants for the life of the lease, typically 25 years.

be used by the mechanical and electrical (M&E) systems designer to specify the plant and also for life cycle calculations. Any external plant, such as sub-stations, are identified. Finally, room data sheets are produced from the user group meetings and tenants requirements, underwritten by the future tenant's representatives. These are a description of each space in the building, what it is for, what specification must the floor, walls etc. have, and what are the fixtures and fittings required. These are accompanied by individual room layouts, which ideally, can be linked by CAD to the room data sheets. Therefore, if any electrical socket is removed from a drawing, it also automatically disappears from the room data sheet. Strangely, in some LIFT contracts this element of the design is not done by the D&C contractor, who regards this as part of the output specification. Other contractors work only from the tenants requirements and schedule of accommodation. This is where the apportionment of risk is important, as discussed earlier:

- If the room data sheets and/or the room layouts are done by others, rather than the D&C contractor, then the risk sits with others too. This may be a healthcare planner (who needs professional *indemnity* insurance to cover his input risk) or another consultant like a space planner. It may be done by the LIFTCo and user groups together, in which case the risk apportionment must be agreed.
- Whatever route is used, the responsibility for adding that output to the architect's design and checking buildability *must* remain with the D&C contractor; otherwise, they can offset risk.
- Or the whole process is given to the D&C Contractor with a careful statement of risks and responsibilities. This may be the most cost-effective route.

At this stage the cost plan is checked, any income from third parties included and the FM costs calculated. The work required to demonstrate value for money from previous schemes or benchmarks, or a new construction tender, is also carried out now. A sense check of the whole project ends this part of the process; if the scheme is drifting into unaffordability, then it can be altered at this stage.

5.2.6 Final scheme design
Alterations will be costly after this stage, when 1:50 drawings are usually provided and a full planning application is lodged on the basis of all the requirements. In this planning process, the use of Section 106 agreements is important. It may be that the new health centre is part of a larger development, in which case the overall developers may not be LIFTCo. Section 106 of the Town and Country Planning Act 1990 in the UK allows a local planning authority to enter into a legally binding agreement or planning obligation, termed as Section 106 agreement, with a land developer over a related issue. Such agreements can cover

almost any relevant issue and can include sums of money. Possible examples of Section 106 agreements are as follows:

- The developer will transfer ownership of an area of woodland to a local planning authority with a suitable fee to cover its future maintenance.
- The local authority will restrict the development of an area of land or permit only specified operations to be carried out on it in the future, for example amenity use.
- The developer will plant a specified number of trees and maintain them for a number of years.

Section 106 agreements can act as the main instrument for placing restrictions on the developers. They often require developers to minimise the impact on the local community and to carry out tasks that will provide community benefits.

During this stage, the detailed design of the structure and required M&E plant are developed. There will also be other specialist input such as infection control staff for instance. To meet the recent UK Building Regulations, thermal modelling of the whole building is required. All applicable British Standards and Health Regulations or guidance, plus the full UK Building Regulations, need to be checked and complied with, or an agreed derogation put in place with the tenants.

Then the final cost is agreed with the D&C contractor, known usually as the guaranteed maximum price (GMP). LIFT contracts have a fixed price, have no variations and are not very flexible, protecting the public sector. As an output from the final tenant requirements and room data sheets, a set of landlords proposals is provided that represents the formal offer to the tenants on which the GMP is based. The LIFTCo also at this stage formally quotes a lease plus payment to the tenants, including all life cycle and maintenance costs.

Once the full planning permission is granted, the LIFTCo will then apply to the local strategic partnering board, and strategic health authority, to proceed to financial close, known as stage 2 approval. In local authority parlance, this is approximately equal to Gateway review 3.

5.2.7 Financial close
The final stage is to put in place the legal documents that will codify the agreement reached in the preceding months. These include, but are not limited to:

- property searches and certificates of title;
- property sales;
- lease plus agreements;
- FM contracts;
- agreed payment mechanisms;
- demised areas in each building;
- D&C contract;

- funders loan agreements;
- shareholder documentation.

It is important to mention the judicial review at this stage – even if planning permission is gained, the general public have 12 weeks to question the legal process that granted that permission. They rarely do so, but if it occurs it stops the project while a judicial review takes place, and the planning decision can be overturned. This is a risk management issue that LIFTCo and its funders must consider if applicable.

The other activity carried out now is financial modelling. It may have occurred earlier and some LIFTCos have it in-house, but in this final stage the tenants will usually expect that an outside agency vets the calculations done by the LIFTCo and all its assumptions (usually estimated in a computer model). Clearly, to run the final model, the D&C contractor confirms his GMP (see earlier).

The funder may also require a separate audit of the model. The tenants will also usually ask the district valuer to evaluate all the property values, and they may also ask for further benchmarking or value for money tests on the LIFTCo proposals. At some point a value-for-money statement will be produced from the health tenants for the Department of Health.

The D&C contractor will then produce the proposed project plan in its final form, with critical path and pinch points shown. It is also usual to agree to the appointment of an independent certifier at this stage. Later on, the certifier will be the assurance for the future tenants that the building conforms to their requirements.

There are also other investors in LIFT, such as Community Health Partnerships, the local councils and private-sector investors, and these are consulted at this stage. Additionally, the strategic partnering board and strategic health authority are required to sanction the financial close. All these bodies will need to pass the project through their boards and gain approval for the scheme.

When all the documents are ready, the parties to the project assemble and commence the financial close, firstly by signing or executing under seal all the agreed documents. The funder will then usually enter the financial markets by telephone and/or electronic linkage, and set up a 25- or 30-year protected funding package, based on the affordability and the financial models run previously. When the most attractive offer is obtained, the final lease plus payment is agreed with the future tenant (it is often slightly different to previous calculations) and the deal is then closed. This commits the tenants to the agreed lease for the whole of its life at a fixed price variable only by the retail price index.

The very next day, the D&C contractor takes possession of the site and starts the set-up for construction.

5.2.8 Construction management set-up
After financial close, there is a period of about 12 weeks while the D&C contractor establishes the site facilities and commences enabling works. If not already

planned and put in place, the developer (LIFTCo in this case) needs to establish its construction management resource. The resource needed is different for each project, so the first task is to assess the complexity and size of the project:

1. Is the site difficult and/or are there many enabling activities before work can commence? (These are sometimes called abnormals, particularly when comparing costs from one project to another.)
2. Is there a lot of M&E work such as ventilation and air conditioning, IT, complex electricals etc.?
3. Is the size of the project such that more construction management resource will be needed?
4. Does the D&C contractor need more managing as a result of past performance, or lack of experience or other factors?

Assuming that the D&C contractor is experienced in PPP work, there should be a lesser need for 'heavy' construction management than is perhaps experienced on other types of contract, as it should not be necessary to issue design or construct instructions like the earlier JCT contracts. However, there is a need for the following:

1. The construction manager (CM) is employed by LIFTCo but is the customer relations manager for the future tenants. The manager needs to ensure that the D&C contractor builds as per the specifications of the landlords' proposals, and that the tenants do not attempt to squeeze in alterations, without realising the implications. Any questions from the tenants go through the CM, who needs to communicate effectively and assuage concerns.
2. There will be monthly evaluations of the work carried out onsite, which need to be verified, checked against the contract and authorised for payment. Some LIFTCos outsource this work to quantity surveyors, others do it in-house, but the activity goes through the CM.
3. The quality of the work onsite has to be regularly checked (the independent certifier will also do this) and the adherence to programme.
4. The health and safety management onsite should be monitored independently from the contractors own staff, and the same should be done for environmental management.
5. All day-to-day contact between the D&C contractor or any other contractors and LIFTCo goes through the CM. Other contractors might include the FM provider for instance.
6. If, unfortunately, problems or variations do occur, then the contract manager will manage the process to resolve them either formally or informally.

As mentioned beforehand, LIFT contracts are typically under £30m in value, so an average project requires the services of a professional CM either full or part time (so for instance it would not be unusual for one CM to handle a tranche of three jobs totalling £30m). As well as holding regular review meetings with contractors and tenants, the CM will produce such formal reports as are necessary for LIFTCo or its customers.

There are, additionally, statutory duties. As in the new CDM Regulations 2007, the developer (LIFTCo in this case) is legally responsible for appointing the CDM coordinator and ensuring that the CDM coordinator has sufficient health and safety expertise to fulfil the duties. The CM therefore leads this process, and it can no longer be devolved to the contractor. It also means that the contractor has to be *actively managed*.

5.2.9 Facilities maintenance

We noted the involvement of the future FM providers earlier with the production of their technical assets assumptions document, but this is only a part of what is needed, as is described in the following paragraphs.

The FM, a key component of LIFT contracts and fundamental to the success or failure of the operation, is of two types, hard and soft FM: the first is concerned with maintaining the fabric and plant, whilst the second relates to the provision of services such as cleaning and waste removal. They are intrinsically linked by what would best be described as operational policies – statements on how the building will be cleaned and maintained, hours of operation, types of activities within etc.

An example of how complicated a situation becomes if this is not thought through at the design stage can be seen in the example of changing a light bulb: The FM Manager of a new health centre is walking through the reception with the Centre Manager and he notices a light fitting high above them in the atrium. 'How do I get to that bulb?' he says. 'Well', says the Centre Manager, 'I phoned the LIFT Company and they phoned the D&C contractor and he said, use a cherry picker'. 'But this reception floor wont take a cherry picker' says the FM Manager, 'because there is an underfloor heating installed' and so on. The final solution is likely that the tower scaffold be assembled specially, with a permit-to-climb and method statement, each time the light bulb fails or investing in a new light that lowers itself.

Given that the FM providers would ask for, and are given, sufficient information to provide their input before financial close, the solution in each case has to be:

- robust and long living;
- cost effective;
- able to prevent the centre being unavailable through equipment failure;
- energy efficient as much as possible, reflecting the latest design and innovation, whilst still being reliable;
- backed by a reactive and proactive service, 24 h, supported by a help desk.

It is not difficult to see that these requirements are a balance. One cannot authorise the latest designs if they are of prohibitive cost and on a whole life basis the extra costs will never be recovered.

As mentioned earlier, the FM has to provide its input not only at the early design stages, so that properly sized plant rooms can be included for instance, but also throughout the project and after the building becomes operational. Later on, the tenants will come to see the hard FM provider as the 'face' of their landlord and the primary contributor to the success of the building. It is up to the LIFTCo, therefore, to monitor this particular supplier closely and install in them a continuous improvement ethic.

5.3 Cultural differences

The most significant cultural barrier to the successful development of a partnering and integrated teamwork approach in the delivery of better value (Thomas and Thomas, 2005). Most issues in the construction process for PPP projects arise from cultural differences, and here we refer to the two cultures of public and private sectors.

The public sector, particularly in healthcare, has risk avoidance as a very high priority, particularly health and cost risks. They are uneasy with LIFTCos, who with a 60% private-sector shareholding are often perceived as private sector in their thinking. Additionally, many primary care trusts who would commission PPP projects have gone through changes from family practitioner committees, to district health authorities, to primary care groups, to primary care trusts; passing through fund holding and several other restructures on the way, all within the last 10 years (1999–2009). During this period much estate and space planning expertise was lost because of revenue budget cut backs and small levels of capital investment in healthcare. Consequently, PCTs were often lacking in the basic resources they needed to plan and commission new facilities. This, and the need to avoid risk, manifests itself occasionally in a difficulty with decision-making on new projects. To the private sector this is an anathema, because they see delay as costing money, and cannot progress their business plans. To the public sector, they must be certain that they can commit large future revenue expenses without affecting their stretched budgets. They must also be assured that the new project will meet the current service needs, and they have to second guess the future requirements. If they cannot see through this mire clearly, they will struggle to make the right decision.

There have been of course, many successful PPP projects through LIFT, and the way through the cultural differences is as follows:

1. Use of the public-sector core team: Most PPPs have a central assurance team independent of the PCTs (they were probably responsible initially for the original LIFT bid). This team monitors and checks the LIFTCo and also brings the projects through the approval stages. Clearly, if the LIFTCo and the core team

work together effectively and transparently (even in the same location), then the hard-pressed PCTs have the reassurance they need. This is also the route to prove value for money on any scheme and/or necessary benchmarking.

2. Demonstrate delivery and cost performance: Trust naturally develops in the partnership when the LIFTCo demonstrates to the partners that when it states a cost, it maintains it, and when it promises a delivery date, the date is met. It may be possible for the LIFTCo to show this by minor capital works projects as well as the larger term jobs.

3. Use of the independent certifier: The appointment of an independent certifier, paid by the project but 'employed' by the LIFTCo, main contractor and public sector, independent of the LIFTCo, is an important assurance for the future tenants. No matter what the main contractor or the LIFTCo says, the certifier will not accept the new building if it does not conform to the tenants requirements. Additionally, if the certifier accepts something, then the contractor and LIFTCo can be confident that they have done this part of the project correctly. A good independent certifier appointment therefore prevents conflict and builds trust.

4. Risk: Going back to 'project set-up', it is worth quoting Smith *et al.* (2006): 'from the viewpoint of risk management, the appraisal phase is the most crucial'. If the project risk is appraised properly at the start and risk apportioned to the partners, then uncertainty is avoided at later stages.

5. Proactive legal teams: LIFT might provide a fertile area for legal practitioners, especially if public and private sectors are apart. To some extent, the Department of Health is now prescribing the documentation to such a degree that local variations are becoming difficult, but the success of a project depends on the following:

    (a) the legal teams on both sides respecting and trusting each other;
    (b) the legal 'heads of terms' at the start of the project are correct; and
    (c) The risk matrix is agreed and understood by all.

    The legal teams are working at the end of the project process, so if a project is formulated badly, it will end up badly, and expensively. A great deal of legal cost is run up by not agreeing a risk strategy and negotiating commercial terms right up to the financial close. Remember, a legal team cannot fix a bad deal – the parties need to involve them earlier in the process to confirm what a *good* deal will look like.

5.4 Conclusions

After reading this chapter, it may appear that PPPs such as LIFT may have struggled to deliver successful projects. The reverse is the case, and at the time of writing, the LIFT process has delivered £1.3bln capital value of new Healthcare buildings in England and Wales, and is arguably the most successful PPP initiative of recent years. The latest initiative known as 'Express LIFT' will enable parts of England and Wales that do not have a LIFTCo to access the LIFT model from pre-approved national companies.

The buildings delivered under LIFT are often landmark public buildings with low energy usage and pleasing, peaceful interiors. Anecdotal evidence suggests that staff retention and patient outcome figures are good. LIFT is still only 4–5 years old, so there is a lack of appropriate measures available. Most importantly, local people are being treated with a raft of services previously not always available outside of specialist clinics or hospitals, local to their homes. To anyone involved with the built environment for healthcare, this is a motivating and worthwhile result.

References

Cartlidge, D. (2006) *Public Private Partnerships in Construction*, Page 10, Taylor and Francis, Oxford.

Smith, N. J., Merna, T., Jobling, P. (2006) *Managing Risk in Construction Projects*, Page 20, Section 2.6, Blackwell Publishing, Oxford.

Thomas, G., Thomas, M. (2005) *Construction Partnering & Integrated Teamworking* 4, Chapter 4, Page 16, Blackwell Publishing, Oxford.

The Integrated Agreement for Lean Project Delivery

William A. Lichtig

6.1 Introduction to Sutter Health

Sutter Health is a healthcare organisation composed of affiliates, which comprise 27 hospitals in Northern California. The primary driver for Sutter Health's lean journey was the California regulatory requirement that healthcare facilities be seismically upgraded to assure capability of continuing in service after an earthquake. This requirement not only created an aggressive time line for completion of their $6bn capital programme over 8 years, but also placed Sutter Health in competition for resources with other healthcare companies obligated to meet the same requirement. A secondary driver was the desire to improve performance in the delivery of capital projects. The manager of facilities development had experienced success with collaborative approaches on projects directly managed. The outside counsel for Sutter Health had a personal connection to the Lean Construction Institute (LCI) (http://www.leanconstruction.org) and brought together Sutter and the Institute's thinking and methods of applying lean to construction.

Sutter Health began with a bang, announcing its commitment to lean delivery of its capital programme on 23 March 2004 in a 150 person meeting with the designers and builders working on its projects. With the help of consultants, they held a series of these meetings devoted to education and discussion. In early 2005, both Sutter Health and its supplier community realised that no one could tell them step-by-step how to do lean construction. However, suppliers understood that Sutter was offering to pay them to experiment with ways of improving performance, and was open to making changes needed to assure supplier profitability under new conditions, roles and responsibilities. Now, 2 years later, architects, consulting engineers, contractors and suppliers meet monthly to share experiences and learnings, even with competitors. Sutter Health has developed a new form of contract, its 'Integrated

Form of Agreement', intended to align interests in pursuit of project objectives. Champions have emerged within the companies involved in the Sutter Health community of suppliers, and in September 2005, persuaded the University of California, Berkeley's Project Production Systems Laboratory (http://P²SL.berkeley.edu) to serve as their learning laboratory. A Prototype Hospital Initiative was launched mid-2006 to develop radically new design concepts for hospitals. The Prototype Hospital Initiative is presented in Chapter 7.

This chapter presents the Sutter Health case study, describing the 'Integrated Form of Agreement' initiative.

6.2 Integrated form of agreement

Traditionally, facility owners have been presented with a standard set of project delivery options: design-bid-build, construction management (agency or at-risk) or design-build. Despite this range of options, many owners remain dissatisfied: projects take too long, they cost too much and the work fails to meet quality expectations. In addition, construction projects continue to present serious safety risks, with nearly 1500 accidents and 4 deaths per day.[1]

Beginning in 2004, Sutter Health embarked on its pursuit of a different project delivery opportunity – one that sought to address the root causes that limited the effectiveness of other models. Sutter Health's delivery model combined 'Lean Project Delivery' and a new contractual model that sought to align the commercial interests of the major project participants and govern the delivery process as a collective enterprise. The 'Integrated Agreement for Lean Project Delivery' offers improved project performance both from the owner's perspective (reduced cost and time, improved quality and safety) and from the viewpoint of the designers and contractors (increased profit and profit velocity, improved safety and employee satisfaction).

6.3 Traditional responses to owner dissatisfaction with the *status quo*

Over the past 100 years, the design and construction industry has become increasingly fragmented. Each specialised participant now tends to work in an isolated silo, with no real integration of the participants' collective wisdom. Sutter Health had adopted many of the common industry responses: post-design

---

[1] According to the US Department of Labor's Injuries, Illnesses and Fatalities Programme, in 2003 there were 1131 fatal accidents and 408 300 non-fatal injuries and illnesses in the construction industry. Website report of US Department of Labor, Bureau of Labor Statistics, Industry at a Glance 'NAICS 23: Construction'. http://www.bls.gov/iag/construction.htm

constructability reviews and value engineering exercises, together with 'partnering' and contractual efforts to shift risk. However, these 'solutions' did not attack the problem at its root cause; rather than working to avoid the problem, providing higher value and less waste, these attempts merely try to mitigate the negative impact of the problems. After a heavy investment of time, money and ego in a proposed design, the inertia against considering the full range of solutions that might be offered by deep value analysis or constructability reviews was routinely quite strong.[2]

Sutter Health concluded that project success required that this fragmentation be addressed directly. This conclusion was supported by a Construction Industry Institute (CII) study determining that the successful projects it studied had certain unifying characteristics. The CII Study concluded as follows:

> Projects are built by people. Research into successful projects has shown that there are several critical keys to success:

1. a knowledgeable, trustworthy and decisive facility owner/developer;
2. a team with relevant experience and chemistry assembled as early as possible, but certainly before 25% of the project design is complete; and
3. a contract that encourages and rewards organisations for behaving as a team (Sanvido and Konchar., 1999).

6.4 What is lean?

The terms 'Lean Production' and 'Lean Manufacturing' largely derive from the Toyota Production System (TPS) (Liker, 2004; Spear and Bowen, 1999; Womack and Jones, 2003). At Toyota, TPS represents only part of a broader business philosophy, known as the Toyota Way. Although a number of 'tools' have been developed that are often identified with TPS, the underlying philosophy and its context are important in this discussion of the use of an Integrated Agreement for Lean Project Delivery. Stated generally, the goal and philosophy of TPS is to produce value, as defined by the customer, without producing waste.

---

[2] The author was recently involved with a project where the mechanical contractor became involved during the Construction Document phase. The project had serious budget problems and the mechanical contractor identified a number of potential cost-saving items. Among those was an idea that would have required some architectural redesign of the penthouse but would save mechanical costs of nearly $800 000. While the entire team basically agreed that the idea was valid, because of the timing and the need for 'redesign' the team seriously considered not pursuing this item. Had the idea been floated during Design Development, there is little doubt that the design would have been modified to achieve this savings without the cost of 'redesign'.

To understand TPS, one first must understand its underlying principles and the context in which it developed. Toyota has its industrial roots as a loom manufacturer. Its initial innovation was to power the looms with a steam engine. Powered looms presented a new dilemma – the loom would continue to run even if the thread broke. Toyota devised a system that would automatically shut down the loom when the thread broke. This system eliminated the waste that would occur if the loom continued running and producing defective material. This principle of 'autonomation' or self-regulation (shutting down production in the face of a defect) was carried forward into TPS as what is often referred to as one of the two pillars of TPS.

Toyota Motor Company was formed in the late 1920s and was only marginally successful. After a visit to the United States, Toyota's chairman challenged the chief engineer to meet US productivity levels (a 10-fold improvement) within 3 years. Toyota did not have the capital, supply chain or infrastructure to support a level of productivity comparable to Ford and GM. Demand for cars in Japan was not constant and consumer demand was more varied. As a result of these limitations, the second pillar of TPS developed: Just-in-Time Delivery. Using 'Just-in-Time Delivery', Toyota only produces items when there is an order; minimising inventories of finished goods. Further, large stores of raw materials or work-in-process is avoided by having those goods 'pulled' to the plant when an order is received. Toyota's ultimate goal is to produce a car to the requirements of a specific customer, deliver it instantly and maintain no inventories or immediate stores.

In order to sustain a system with no inventory or work-in-process, Toyota needed to produce items without defects because a defect would require stopping the production line. Further, this would require tight coordination between all sections of the factory, using clear language and systematically requesting parts and materials at the proper time. To assure that defective parts are not forced further into production, Toyota workers are now expected to act as the autonomic loom had: stopping production if they find a defect. This system decentralises authority and empowers factory workers in ways that were previously unprecedented in the West (or in Japan, for that matter).

Finally, the implementation of Just-in-Time Delivery shifted the focus from the productivity of each unit in the factory to the overall productivity of the system. Because no unit could individually produce parts or perform a function to create inventory, units were only as productive as the overall system. This had the benefit of keeping the entire factory focused on 'through-put', the output of the entire plant.

6.5 The application of TPS principles to design and construction

The LCI has grappled with how the ideals of TPS and lean production could be applied to design and construction. One of the fundamental differences identified in the project setting was that design and construction are not a high-volume, repetitive process like car manufacturing. What manufacturing accomplishes by the arrangement of the factory or modification of its machines cannot be replicated in design and construction.

Greg Howell and Glenn Ballard, the founders of the LCI, hit the proverbial nail on the head. They identified the key 'item' flowing on a project is the work that is completed by one performer and handed-off to his successor. Like Just-in-Time deliveries of materials, what was being delivered was work from one trade to another. Finding that the theories of dependence and variation[3] largely explain what happens on a project when reliable workflow between trades is not maintained, they developed a planning system that enables a project team to focus its attention on causing work to flow across the value stream (Howell, 1999). This system also incorporated the idea of autonomation by distributing responsibility for developing and maintaining the planning model to the 'last planners' – the individuals who, like the factory workers at Toyota, need to be in a position to stop production if the preceding work is defective.

Needless to say, the Last Planner System™ (LPS) created quite a stir in the 'command and control' structure embedded in construction project management since the days of Fredrick Taylor. Its goal is to create reliable workflow by having the project team, including all affected firms, collaboratively create a phase plan for a segment of the work (e.g. foundations). Thereafter, a 6-week look-ahead plan is prepared where the team identifies the constraints or prerequisites that must be satisfied for a work assignment not to be 'defective'. Each week, the team screens upcoming work assignments for 'defects' (e.g. unanswered Requests For Information (RFIs), incomplete prerequisite work, missing materials, lack of equipment or labour resources) and only releases work to the field that has no constraints. Work commitments are then obtained, again from the 'last planners', based upon requests made to the last planners. Because the system values reliability over speed, 'last planners' are *expected to decline* an assignment if either the assignment is defective or if they lack confidence that they actually can perform the task. By saying 'no', the team is then able to re-plan the work and avoid the waste that is created when the downstream performer plans staffing, deliveries etc., based upon a flow that will not happen.

At its core, 'the essential work of projects is conducted as conversations . . .' and 'the work of business in making and keeping commitments' (Macomber and Howell, 2003). As such, the effectiveness of the LPS is dependent upon the concept of making

---

[3] The impact of dependence and variation on flow in the manufacturing setting was described in *The Goal* by Eliyahu Goldratt and became the basis for the Theory of Constraints. *See* Focused Performance website, http://www.focusedperformance.com/toc01.html

[4] This articulation is derived from the Linguistic-Action model and the work, among others, of Fernando Flores. These ideas have been brought to the Lean Construction community by Hal Macomber, a principal in Lean Project Consulting, Inc. A full discussion of Linguistic Action and its relationship to LPS can be found in Macomber and Howell (2003).

and securing reliable promises.[4] Before making a commitment, then, the last planner must (1) determine that the proposed performer is competent; (2) estimate the time to perform; (3) confirm availability and allocate capacity for the estimated time; (4) assure there are no hidden doubts about performing; and (5) be prepared to be responsible for any failure. Obviously, to make these assessments, the last planner may often need to be in conversation with others and seek their commitments as well. Hence, as discussed below, projects become networks of commitments.

Proper use of LPS produces reliable workflow and stabilises the project. It results in reduced costs, shortened durations, increased quality and increased safety. The LPS is to the project setting what Just-in-Time Delivery is to manufacturing. With TPS, 'lowering the "water level" of inventory exposes problems (likes rocks in the water) and you have to deal with the problems or sink. Creating flow, whether of materials or of information, lowers the water level and exposes inefficiencies that demand immediate solutions' (Liker, 2004, p. 88). Employing the LPS produces stable workflow, allowing the project team to explore other opportunities to eliminate waste from the design and construction process. Some of these other opportunities – such as Target Value Design and Built-In Quality – are described below when the terms of the Integrated Agreement are discussed.

6.6 Sutter Health's formulation of a lean project delivery strategy

Sutter Health's approach to Lean Project Delivery strives to coherently address each level of the project delivery system – the physics of work, organisational structure and commercial relationships. This approach has become known in the Lean Project Delivery community as the Five Big Ideas. The Five Big Ideas are summarised in Figure 6.1.

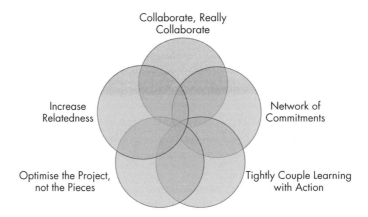

Figure 6.1   The Five Big Ideas.

The Five Big Ideas form the framework for approaching all aspects of Sutter Health's Lean Project Delivery. The description that follows is taken from the manifesto that was signed by members of its Facility, Planning & Development (FPD) and its design and construction community at the outset of Sutter Health's lean initiative:[5]

1. Collaborate; really collaborate, throughout design, planning and execution.
   Constructable, maintainable and affordable design requires the participation of the range of project performers and constituencies. Because abandoning the master-builder concept and separating design from construction, we have been patching a poorly conceived design practice. Value engineering, design assist and constructability reviews mask an underlying assumption — that design can be successful when separated from engineering and construction. Design is an iterative conversation; the choice of ends affects means, and available means affects ends. Collaborative design and planning maximises positive iterations and reduces negative iterations.
2. Increase relatedness among all project participants.
   People come together on AEC projects as strangers. They too often leave as enemies. Healthcare facilities projects are complex and long-lived, requiring ongoing learning, innovation and collaboration to be successful. The chief impediment to transforming the design and delivery of capital projects is an insufficient relatedness of project participants. Participants need to develop relationships founded on trust if they are to share their mistakes as learning opportunities for their project and all the other projects. This will not just happen. However, we are learning that relationships can be developed intentionally.
3. Projects are networks of commitments.
   Projects are not processes. They are not value streams. The work of management in project environments is the ongoing articulation and activation of unique networks of commitment. The work of leaders is bringing coherence to the network of commitments in the face of the uncertain future and co-creating the future with project participants. This contrasts with the common sense understanding that limits planning as predicting, managing as controlling and leadership as setting direction.
4. Optimise the project not the pieces.
   Project work is messy. Projects get messier and spin out of control when contracts and project practices push every activity manager to press for speed and lowest cost. Pushing for high productivity at the task level may maximise local performance but it reduces the predictable release of work downstream, increases project durations, complicates coordination and reduces trust. In design, we incur rework and delays. In the field, this means

---

[5] The Five Big Ideas and the resulting manifesto were developed by Lean Project Consulting, Inc.

greater danger. We have a significant opportunity and responsibility to reduce workers' exposure to hazards on construction projects. Doing so can bring about greater than 50% improvements in the safety on the work site. As the leading community-based healthcare system in northern California we are committed to do all that is possible so that the people who build these projects are able to go home each night the way they came to work. The way we understand work and manage planning can increase that messiness or reduce it.

5. Tightly couple action with learning.
   Continuous improvement of costs, schedule and overall project value is possible when project performers learn in action. Work can be performed in a way that the performer gets immediate feedback on how well it matched the intended conditions of satisfaction. Doing work as *single-piece flow* avoids producing batches that in some way do not meet customer expectations. The current separation of planning, execution and control contributes to poor project performance and to declining expectations of what is possible.

6.7 Development of the integrated agreement for lean project delivery

In order to fully embrace the Five Big Ideas, Sutter Health determined that it should develop a relational contract[6] – an agreement that would be signed by the architect, the construction manager/general contractor (CM/GC), and owner – and would describe how they were to relate throughout the life of the project. Further, the new relational agreement would also adopt the underlying principles of Lean Project Delivery and the Five Big Ideas the projects operating system so that all members of the 'Integrated Project Delivery Team' would have a clear understanding of how the project would be administered. What follows is an 'executive summary' of some of the major elements of Sutter Health's Integrated Agreement for Lean Project Delivery organised around the concepts of the Five Big Ideas.

6.7.1 Relationship of the parties

The Integrated Agreement is a single contract signed both by the architect and the CM/GC. It is not a design-build agreement, where one entity takes total responsibility for all aspects of project delivery. Instead, the Integrated Agreement describes the relationships that are established among the members of the Integrated Project Delivery (IPD) Team, recognising that different members, whether traditionally

---

[6]The concept of relational contracts (as opposed to transactional contracts) was developed by Ian Macneil. For a historical retrospective of McNeil's theory of relational contracts, *see* Campbell (2004).

a consultant or a subcontractor, may have design responsibility. From the outset, the Integrated Agreement seeks to create coherence between the interests of the project and the participants and to align the interests of the project performers. The Integrated Agreement calls for team members to be selected based upon responses to Requests for Proposal – it is a quality, value-based selection rather than based upon lowest price. Conceptually, the primary members of the team are selected at the outset of the project. Whether the architect or CM/GC is selected first, or whether a group comes forward as a self-assembled team depends largely on the preference of the owner. However, since historically the architect has been selected first, an interesting message of commitment to change can be signalled by selecting the CM/GC first.

The direct parties to the Integrated Agreement are the owner, the architect and the CM/GC. Rather than being conceived as a 'three legged stool', this primary relationship is depicted as three overlapping circles. The project representatives for each of these entities form the 'Core Group'. This group, which may also invite other members of the IPD Team to join (or leave) the Core Group, has primary responsibility for the selection of the rest of the IPD Team and for management and operation of the project. Most major project-related decisions are to be made by consensus of the Core Group. Only in the event of impasse does resolution of issues transfer to the owner. The Core Group, which is to meet regularly, is also responsible for developing and implementing various project plans that reflect the Core Group's strategy for communication, planning, quality and other aspects of the project.

The Core Group is also responsible for joint selection of other members of the IPD Team. While the owner, CM/GC and architect each can recommend firms from whom proposals should be solicited; ultimately the list is developed and approved by the Core Group. Once additional IPD Team members are chosen, each is expected to sign a joining agreement, acknowledging that the firm is familiar with the terms of the Integrated Agreement and agrees to participate in the project based upon the described level of responsibility and collaboration. To facilitate integration into the team and the anticipated level of collaboration, the Integrated Agreement contemplates that the major consultants and subcontractors will be selected during the validation phase (see discussion below). By bringing the team together early, the agreement seeks to gain maximum participation and innovation when the team's efforts are likely to have the greatest financial impact (Sanvido and Konchar, 1999).

The Integrated Agreement also calls for executive oversight for the Core Group to foster learning and a collaborative environment. Senior executive representatives are expected to join the Core Group meetings on at least a quarterly basis. In addition, the senior executives are expected to participate in problem solving in the event the Core Group is unable to promptly resolve an issue. Similarly, in addition to the Core Group meetings, the Core Group is called upon to schedule regular IPD Team meetings to address project design and construction issues, to confirm that information is being shared across project teams, and to gain the benefit of having shared expertise to address pre-construction issues.[7]

Finally, the Integrated Agreement expressly sets forth the goals of forming an IPD Team:

> By forming an Integrated Team, the parties intend to gain the benefit of an open and creative learning environment, where team members are encouraged to share ideas freely in an atmosphere of mutual respect and tolerance. Team Members shall work together and individually to achieve transparent and cooperative exchange of information in all matters relating to the Project, and to share ideas for improving Project Delivery as contemplated in the Project Evaluation Criteria. Team members shall actively promote harmony, collaboration and cooperation among all entities performing on the Project.

The parties recognise that each of their opportunities to succeed on the Project is directly tied to the performance of other Project participants. The parties shall therefore work together in the spirit of cooperation, collaboration and mutual respect for the benefit of the Project, and within the limits of their professional expertise and abilities. Throughout the Project, the parties shall use their best efforts to perform the work in an expeditious and economical manner consistent with the interests of the Project.

6.7.2 Creating a collaborative design and construction environment

Collaboration occurs best when the participants view themselves as equal in the process and when the initial collaboration centres on exploring and defining the problem, rather than commenting on another's proposed solution. The Integrated Agreement recognises this need as follows:

> In order to achieve owner's basic value proposition, design of the Project must proceed with informed, accurate information concerning program, quality, cost and schedule. While each IPD Team Member will bring different expertise to each of these issues, all of these issues and the full weight of the entire teams' expertise will need to be integrated throughout the pre-construction process if the value proposition is to be attained. None of the parties can proceed in isolation from the others; there must be deep collaboration and continuous flow of information.

---

[7]Toyota has used what it refers to as 'obeya' or ' big room' meetings to gain the synergy that develops when cross-functional teams are brought together under one roof to explore problems. As noted by Toyota executive Takeshi Yoshida, 'There are no taboos in oobeya. Everyone in that room is an expert. They all have a part to play in building the car. With everyone being equally important to the process, we don't confine ourselves to just one way of thinking our way out of a problem' (Warner, 2002).

In support of the goal to make the owner's value proposition paramount, the Integrated Agreement calls for the Core Group to develop a Target Value Design[8] plan and requires the IPD Team members to provide Target Value Design support services throughout development of the design. Target Value Design is intended to make explicit that value, cost, schedule and constructability (including work structuring) are basic components of the design criteria. It contemplates that the owner will have a series of value propositions (e.g. a desire that each worker have access to natural light), in addition to its purely programmatic needs, which may need to be ranked to achieve the basic business case. The Core Group's Target Value Design plan is expected to address formation and meeting schedule of cross-functional teams or clusters; meetings for the system or cluster leaders to share information about their system with those responsible for other systems; continuous cost model updating to assure that on-going design is not exceeding budget; and methods for evaluating Target Value Design tradeoffs and opportunities (including function/cost trade-offs) to maintain total project target cost.

The goal of Target Value Design is to enable the design to proceed informed, on a real-time basis, by the cost, quality, schedule and constructability implications of proceeding with a design concept. Traditionally, the construction team participated, if at all, only after designs have been committed to paper and thrown over the wall – performing 'un-constructability analysis' and 'de-value engineering'. At best, this resulted in negative iteration and waste when designs had to be changed when they proved to be over budget or not constructable. Instead, the Integrated Agreement seeks to create the equivalent of 'paired programming', where individuals with different backgrounds and expertise simultaneously, side-by-side, attack the same problem, allowing each to benefit from the expertise of the other. The team is expected to engage in design reviews with an eye towards value – constantly exploring whether other construction options will better serve the owner's value proposition.

The Integrated Agreement also permits the Core Group to identify which firm will have design responsibility for a given scope. It expects that major portions of the project will garner the participation of design-collaboration or design-build subcontractors (mechanical, electrical, plumbing, fire, curtain wall and skin). Again, the design process is structured to encourage the sharing of intermediate design documents, rather than just handing off large batches of drawings at extended intervals.

The Integrated Agreement also expects that the Core Group will collaboratively develop a joint site/existing condition investigation plan, proposing the level of investigation that the team recommends as prudent. In addition, the Core Group jointly develops the scope for third-party consultants and collectively assesses the

---

[8] Target Value Design is similar to Target Costing, but may be broadened to encompass additional design criteria beyond cost, including time, work structuring, buildability and similar issues. For a discussion of Target Costing *see* Ballard and Reiser (2004).

resulting work product to evaluate it for completeness and sufficiency to inform design and construction.

Collaboration does not end when the contract documents are approved for construction. The Integrated Agreement also calls for the Core Group to develop a Built-In Quality Plan. Although reports vary, it is estimated that up to *10% of project construction cost* is spent on field re-work.[9] The goal of the Built-In Quality Plan is to cause the IPD Team to openly develop ways to ensure that the expectations of the firms and individuals who will be responsible for accepting the work are communicated to the workers who will be executing the work. In addition, the plan should empower workers to 'stop the production line' if they determine that work is being passed along that does not meet the agreed-upon hand-off criteria. Again, the overall goal is for all project participants to collaborate in advance about what is required and put systems in place to 'mistake-proof' the process and minimise the amount of re-work.

Another example of focused collaboration is in the realm of problem solving and dispute resolution. Initially, problem solving is facilitated by the Core Group. Rather than making the architect the arbiter of project disagreements, the Integrated Agreement calls upon the Core Group to conciliate and resolve these issues. If they are unable to do so, then the Senior Management representatives are expected to join the Core Group in a meeting to resolve the issue. If the issue is still not resolved, the Core Group may elect to retain an independent expert to review the issue and provide an unbiased assessment to the Core Group. Each of these levels is an effort to allow the team the opportunity to resolve any issues without creating direct adversity where one among a group of equals is empowered to make a 'decision'.

6.7.3 Articulating and activating the network of commitments

The Integrated Agreement acknowledges that the ability to establish reliable workflow is dependent on the making and securing of reliable promises.

Fundamental to the success of Lean Project Delivery is the willingness and ability of all IPD Team members to make and secure reliable promises as the basis for planning and executing the Project. In order for a promise to be reliable, the following elements must be present:

- the conditions of satisfaction are clear to both parties – the performer and the customer;
- the performer/promisor is competent to perform the task or has access to the competence and the wherewithal (materials, tools, equipment, instructions) to perform the task;

---

[9]The CII's study entitled 'Costs of Quality Deviations in Design and Construction', (Pub 10-1) concluded that the average re-work on industrial projects exceeds 12%, equating to waste of $17bn annually. *See also* Construction Owner's Association of Alberta, Project Rework Reduction Tool available at http://rework.coaa.ab.ca/library/prrt/default.htm

- the performer/promisor has estimated the time to perform the task and has internally allocated adequate resources and has blocked the time on its internal schedule;
- the performer/promisor is sincere in the moment that the promise is made – only making the promise if there is no current basis for believing that the promise cannot or will not be fulfilled; and
- the performer/promisor is prepared to accept the legal and reasonable consequences that may ensue if the promise cannot be performed as promised and will promptly advise the IPD Team if confidence is lost that the task can be performed as promised.

One area where the Integrated Agreement seeks to implement the Linguistic Action model, focusing on requests and promises, is concerning RFIs. Under the traditional model, an RFI is often submitted, logged, tracked, hot listed and ultimately responded to without any direct conversation between the parties, without regard to the work activity affected by the RFI, and without any promise, reliable or otherwise, being made about when a response might be forthcoming. The Integrated Agreement first makes the bold stand of stating a 'zero RFI goal' given the deep level of pre-construction collaboration. In the event that clarification is needed, however, the agreement provides:

> To the extent that the need for clarification does arise, the party seeking clarification should first raise the issue either in a face-to-face conversation or via telephone in accordance with the Project Communication Protocols. The initial conversation shall describe the issue, identify the area affected and request the clarification needed. If the parties to that conversation are able to resolve the issue in the course of that conversation, they shall also agree on how the clarification shall be documented and reported to the Core Group. If the parties to that conversation are not able to resolve the issue in the course of that conversation, they shall agree on how the issue will be resolved (who, will do what, by when) and shall agree which of them will notify the Core Group concerning the issue and how they plan to resolve it. It is the parties' goal that RFI's will only be issued to document solutions, rather than raise questions that have not previously been the subject of a conversation. To the extent that resolution of the issue may affect progress of the Work, the issue shall be included in the planning system.

The Integrated Agreement also calls for the project planning system to be based on collaborative, pull planning – using the LPS or an equivalent. It identifies the fundamental characteristics that must be met:

> At a minimum the system must include a milestone schedule, collaboratively created phase schedules, "make-ready" look ahead plans, weekly work plans, and a method for measuring, recording, and improving planning reliability.

The Integrated Agreement goes on to describe each of these elements in further detail and what is required at each level of the planning system. It also describes the elements of the planning system that need to be addressed at the 'Weekly Look Ahead Planning Meeting' (identification and promises for removal of constraints – e.g. RFI responses which must precede identified work) and the 'Weekly Work Planning Meeting' (reliable promises from last planners of what work identified in the Look Ahead Process as constraint-free, will be completed to agreed-upon, hand-off criteria each day and by week's end). Finally, the system must capture and calculate planning reliability and root causes for variance so that the IPD Team can develop a plan to improve reliability.

6.7.4   Optimising the project, not the pieces

The Integrated Agreement seeks to create a system of shared risk, with the goal of reducing overall project risk, rather than just shifting it. In part, this goal is supported by investing significant efforts in up-front collaboration, with the owner funding early involvement of the project team in an effort to eliminate ambiguity in the documents and maximise the collective understanding of the project's conditions of satisfaction. The Integrated Agreement also strives to raise the quality of design by insisting that design fees be supported by a resource-loaded work plan. The CM/GC is compensated on a cost-plus fee basis with either a guaranteed maximum price (GMP) or an estimated maximum price (EMP). An EMP operates as a pain and gain sharing threshold, but limits the potential loses to the IPD Team at their collective profit, keeping with the owner the risk of more significant cost overruns. Some subcontractors are also compensated on a cost-plus basis. GMP/EMP proposals usually are based on drawings submitted for permit, reducing the need for added contingency.

Historically, project owners have established separate contingency amounts for design issues and construction issues. The Integrated Agreement combines these contingencies into one IPD Team contingency. The benefit of this shared contingency is that it focuses each team member not only on its own performance, but also on the quality of other team member's performance, as well. In this way, the success of every team member is directly tied to the performance of all members of the IPD Team. Furthermore, access to contingencies is jointly managed throughout design and construction by the Core Group. The sharing of contingency begins to shape the sense of a collective enterprise.

In addition, as a result of their early involvement, the CM/GC and trade contractors agree to a limited basis for change orders – material scope changes, changed site conditions, or unforeseen regulatory or code interpretations. The traditional bases for many change orders – lack of document or discipline coordination – are eliminated as a result of the coordination efforts during the design phase. Despite its lean ideals, the Integrated Agreement does not contemplate perfection; the IPD Team Contingency is made available to address work that was inadvertently omitted from the GMP/EMP estimate or results from coordination mistakes.

The Integrated Agreement also eliminates the traditional 'negligence' standard as the measure of the designers' financial responsibility. Instead, the owner and the Core Group members agree that the IPD Team Contingency can be used to cover construction costs for 'errors & omissions', even those resulting from negligence. While the designer would still have access to insurance for costs that resulted from work that fell below the standard of care, this would not be the exclusive recourse. This system allows the parties to establish an agreed level of quality and share the risk without being forced into an adversarial system that creates significant waste. With the level of quality established, the architect is able to prepare its resource loaded work plan accordingly.

In the past, some owners have used a 'shared savings' mechanism; however, this may cause optimising the pieces and forecloses participation of the design team. The Integrated Agreement permits the Core Group to adopt an incentive sharing plan 'to encourage superior performance' based upon the Lean Project Delivery goals. The programme must be fashioned to support the Five Big Ideas and balance between the different behaviours and results called for by those concepts. Any programme is expected to consider performance in the following areas: cost, quality, safety, schedule, planning system reliability, and innovative design or construction processes. The programme must provide a basis for establishing project expectations and benchmarks and continually monitoring and reviewing the project team's performance, providing the team with periodic performance information to allow corrections or modifications *during* project performance to improve the quality of the services provided. Also, the team must participate in the pool so that it supports the creation of one, unified team focused on overall project performance. Again, this enhances the sense of a collective enterprise.

The incentive programme would be funded with project savings as evidenced by both contingency preservation and reduction in the project's Costs of the Work as compared to the amounts contained within the GMP. These savings would create the 'incentive pool' which would then be paid, based upon evaluation of performance against other performance criteria. For example, the Core Group might establish performance goals in at least the following areas: quality, safety, planning system reliability and innovative design or construction processes. The team's goals would be expressed as a range of outcomes from 'business-as-usual', to 'stretch goals', to 'exceptional performance'. Performance would be monitored and rated, with the overall portion of the incentive pool to be paid to the team based upon performance on the non-cost performance criteria.

6.7.5 Tightly couple learning with action

Too often, projects are completed without capturing the learning; 'lessons learned' are discussed at project completion to be applied on the 'next' project. One of the Five Big Ideas is to 'Tightly couple learning with action'. If periodic project reviews are not performed, then the opportunity for improvement over the life of a multi-year project is lost. Moreover, the existence of financial incentives provides added motivation for individuals and organisations to stretch beyond their current levels of performance or ways of doing business and may help overcome the inertia and resignation that often exists on projects.

The concepts of continuous improvement and learning from project performance are embedded in many of the Integrated Agreement's performance requirements. As discussed above, the planning system calls for weekly assessments of planning system reliability and reasons for variance, with the IPD Team responsible for determining ways to reduce variability. Similarly, monthly assessments are to be made during construction of root causes of contingency utilisation and change orders with the goal of minimising future need.

The Core Group is specifically charged with developing the project evaluation criteria (this may be performed in conjunction with the Incentive Sharing Plan), conducting periodic project assessments, and planning and implementing 'programmes to improve Project performance and performer satisfaction with the Project'. Similarly, the Built-In Quality Plan specifically must address how to assess performance, identify root causes and continuously improve performance.

6.8 Conclusion

The Integrated Agreement for Lean Project Delivery is a significant departure from other project delivery and contractual models. It seeks to align the commercial relationships with the lean ideals. It also recognises the highly relational nature of the interactions of a construction project's design and construction participants that are assembled as a temporary production system. Buying design and construction is not like buying a commodity. The Integrated Agreement has been developed in an effort to support the values of Lean Project Delivery that are exemplified in the TPS – the elimination of system-wide waste and the pursuit of value from the owner's perspective. Rather than focusing on risk transfer, the Integrated Agreement seeks to establish systems and empower the IPD Team to reduce or eliminate risk by employing new conceptual and autonomic approaches to project delivery. Early assessments of Lean Project Delivery support the conclusion that risks associated with time, cost, quality and safety issues can be reduced by implementing lean thinking. The Integrated Agreement should support deepening those efforts and further reduce those risks.

References

Ballard, Glenn; Reiser, Paul (2004). The St. Olaf College Fieldhouse Project: A Case Study in Designing to Target Cost. Proceedings of the 12th Annual Conference of the International Group for Lean Construction, available at http://www.iglc.net/

Campbell, David (2004). *Ian Macneil and the Relational Theory of Contract*. Center for Legal Dynamics of Advanced Market Societies, Kobe University.

Howell, Gregory A. (1999) What is Lean Construction. Proceedings, Seventh Annual Conference of the International Group for Lean Construction, available at http://www.iglc.net/

Liker, Jeffrey K. (2004) *The Toyota Way: 14 Management Principles from the Worlds Greatest Manufacturers*. New York, USA.

Macomber, Hal; Howell, Gregory (2003) Linguistic Action: Contributing to the Theory of Lean Construction. Proceedings of the 11th Annual Conference of the International Group for Lean Construction, available at http://www.iglc.net/

Sanvido, Victor E.; Konchar, Mark D. (1999) *Selecting Project Delivery Systems: Comparing Design-Build, Design-Bid-Build and Construction Management at Risk*. State College, PA: The Project Delivery Institute, 1–77.

Spear, Steven; Bowen, H. Kent (1999) Decoding the DNA of the Toyota Production System. *Harvard Business Review*, vol. 77, no. 5, p. 96 (Sept/Oct 1999).

Warner, Fara (2002) In a Word, Toyota Drives Innovation. Fast Company, August 2002, available at http://www.fastcompany.com/magazine/61/toyota.html

Womack, James P.; Jones, Daniel T. (2003) *Lean Thinking: Banish Waste and Create Wealth in Your Corporation* (2nd edn). London: Simon & Schuster UK Ltd.

The Sutter Health Prototype Hospital Initiative

Dave Chambers

This chapter presents a continuation of the Sutter Health case study, presented in Chapter 6. Please see the introduction of Chapter 6 for a description of Sutter Health and the history behind the Prototype Hospital Initiative.

There are underway, not least in Sutter Health, impressive efforts to transform the way healthcare facility projects are delivered (see the cover story 'Lean but not Mean', ENR, 24 November 2007). Yet even if project delivery is dramatically improved, from the perspective of healthcare operations, will the project positively impact the services provided?

This is much more than a rhetorical question about patient care. If we are motivated to rethink project delivery, it is as a result of the conundrums resulting from the fragmentation and gaps in communication characteristic of the current design and construction models. Whether projects are delivered as design-build or design-bid-build, negotiated or competitively bid, in my experience, our problems reside in the non-existent alignment of participants' incentives, very weak networks of commitments and an overall failure to collaborate. In healthcare, this is true of the workings within the facility as well.

The formation of departments with their own highly focused and very limited areas of accountability and output measures (generally disassociated with other departments) has created a very similar set of conundrums. Departments are measured as cost centres. Each is responsible for its own optimisation, not the optimisation of other departments or even the ultimate output: a successful patient discharge.

Yet, the discipline of architecture is concerned with the design of spaces in which human beings dwell, work and encounter each other and themselves. Each architectural design encourages particular ways of being, disciplines, activities and human enterprise, and discourages others. Each design opens particular possibilities and features of life, and closes others. In other words, the system of optimised pieces that the design and building industry has developed has failed on virtually every level of the enterprise.

To address this problem from a holistic perspective, Sutter Health engaged in a unique process of building a knowledge base for project planning, design and construction parameters. The development of these parameters was all to be based on lean concepts. The development process itself was also uniquely founded upon lean philosophy. This was the Prototype Hospital Initiative.

Sutter Health identified the fundamental motivation for the project as the need to dramatically improve patient safety, clinical and logistic operations efficiencies, construction cost containment and facility sustainability. The prototype would be a scalable 60-90-120 bed secondary care, general acute services facility. This was to be a Greenfield site and the proposed solutions would be adaptable to varying site constraints. Adaptability in this case was more focused on building design features and standardisation of the key processes (clinical and logistic) than on a specific floor plan.

7.1 Getting started

To begin this process, Sutter requested that participants self-select to provide a complete core team to deliver the project requested. Since Sutter fully intended to challenge the essential workflows characteristic of contemporary acute care environments, the core team also included clinical operations consultants. Sutter reasoned, if space tends to force workflow compromises, then the paradigm could change if the spatial plan was informed by optimised concepts of workflow. One of the required deliverables for the engagement defined in the request for proposals was an actual staffing model for the designed solution. In effect, if 'form follows function' (Louis Henri Sullivan) then if the function, or physical process is optimised, then the result should be that fewer staffing resources would be required for 'like' services. To start this process, the teams were given comparative basic workflows intended to catalyse the discussion. Figure 7.1 provides an example of the comparative differences in workflow from a traditional departmentalised Diagnosis and Treatment (D&T) workflow typical of many acute care facilities to an optimised workflow based on minimised patient movement. (Patient movement was considered to be waste.) The number of discreet steps in the provision of care articulated as the current state (diagram on the left in Figure 7.1) represents over 66% more steps in the delivery of most key services in this diagram versus the optimised state shown on the right.

In addition to the clinical operations expertise, teams included essential project planning, programming, design and construction expertise. Teams were asked to think of themselves as a single project delivery entity. The request for proposals identified an even number of architects and construction managers and also identified other possible team members in the Request for Proposal. Teams

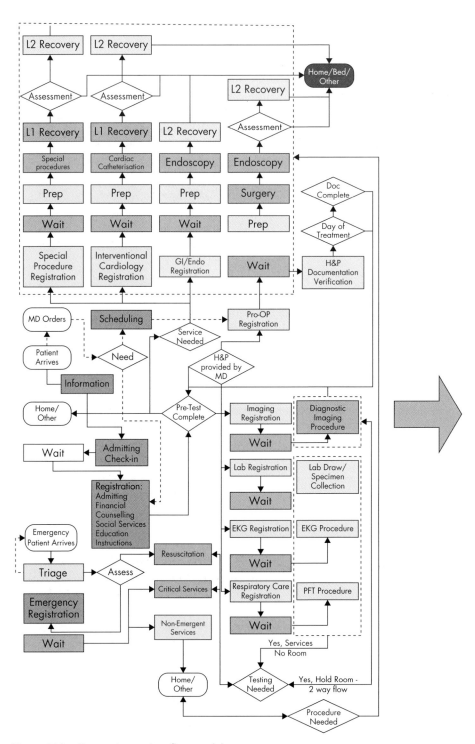

Figure 7.1 Contrasting patient flow models.

were asked to envision a structure that would strongly emphasise interactive communication and creative intersectional thinking. Sutter Health was, in effect, inviting the intersections of different disciplines. It was up to the teams to demonstrate how those intersections would inform their process. The teams' performance would be measured by key metrics based on operational efficiency, programme efficiency (area), building costs (first and lifecycle), project time to complete and facility energy consumption. They would also be evaluated on their willingness to collaborate, not only within their team, but even with other participants.

As teams formed their responses, Sutter Health formed a steering committee comprised of hospital executives, physicians, nurses, architects, engineers and planners. The purpose of this team was not only to select the teams but to engage with them to provide input and ultimately to evaluate the strength of their proposed solutions.

In order to accelerate learning, the format of the engagement for the Prototype Hospital Initiative was to be a 'co-opetition'. Not one, but three fully self-assembled teams were to participate in the development of an early schematic

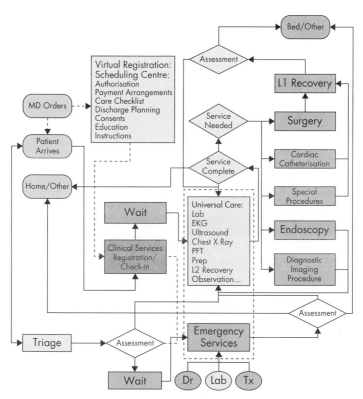

Figure 7.1 (Continued)

package sufficiently developed to define the project to be actually developed on the first site in terms of spatial programme, layout of services, building systems and costs, and energy requirements for the enclosed envelope. Each team selected to participate in this co-opetition was paid a stipend of $500 000 for their participation.

Three teams, three designs, one set of defined parameters: this was a unique approach to true set-based design (originally proposed by Ward *et al.*, 1995). At the outset, in order to foster deep collaboration, Sutter representatives established that due to overall demand, all teams would be provided the opportunity to advance into project delivery on one of the prototype hospitals to be built. The strength of their effort as measured by the criteria established in this section would determine the sequence of project awards to the teams.

7.2 Goals and metrics

Once the three teams were selected, Sutter Health convened the first of several charrettes for the teams to establish clear measures of success, and a consensus-driven formal project schedule, including identification of all interactive charrettes and ending with the submission of each team's final deliverables for the co-opetition phase of the project. This charrette was held in mid-August 2006.

The conversations were dynamic and stimulating. Objectives established for the project were aggressive, outside of participants' comfort zones but not seen as completely unattainable or unreasonable. Sutter noted that leapfrog change is not a refinement of current performance. It is by definition, aggressive and challenges the *status quo*. The objective include:

- 40% improvement in workflow efficiencies as measured by discreet steps in the overall care process, aggregated cycle times for key outcomes, and resource use per adjusted patient discharge (APD) (as modified by case mix index);
- 30% decrease from baseline area required per APD;
- 50% decrease in time to build;
- 25% decrease in energy consumption per APD.

The important criteria to note in these metrics are APDs, or throughput as the baseline for these measured improvements. Throughput is a capacity of service completely independent of revenue and provides the opportunity to simply evaluate the efficiency of a process. These metrics led to the beginning of the project teams' work to redefine clinical workflow. Before allowing designers to begin designing, clinical workflows based on patient movement were studied with the notion of optimising these flows before any design or even spatial programming could begin.

Teams at first found this somewhat challenging. Architects are used to programming and designing out of the gate and contractors are used to looking at drawings to establish costs. Both disciplines were now being asked to learn the workflows to be resolved and to use what they eventually would learn to rethink how to approach their work. What came out of this effort was development of the multidisciplinary service 'cell' in lieu of departmental constructs of service.

The cellular concept is not new to manufacturing but, in my experience, is virtually undiscovered territory in the world of healthcare. Whereas departments, as described earlier, measure narrowly defined outputs against cost, cells measure 'milestone' outputs in the care continuum. Cells are configured, with all disciplines required to complete key steps in the care delivery process thereby minimising handoffs, discreet steps of movement and cycle times in the care delivery process. For example, the intake cell conjoins admitting, lab, electrocardiogram, health screening and pulmonary function departmental services in a single cell. The output of the cell is that the patient is correctly assessed for the treatment or acute care pathway he/she needs and is ready for that treatment or acute care pathway.

Similarly, output of a treatment cell (surgery, endoscopy, cardiac catheterisation and radiographic special procedures) is that the patient is successfully discharged or is successfully admitted to the appropriate acute care unit. Output of the acute care unit is a successful discharge. Diagnostics are typically decentralised to the cells that they would act in to support these outputs.

7.3 Design

After two interactive charrettes (charrettes two and three) exploring and developing these concepts over the course of nearly 3 months, the project teams advanced into facility programming, design and building system evaluation.

The fourth charrette facilitated in mid-November 2006 explored architectural interpretations of cellular flow. The big idea that emerged from this effort was the development of a 'universal care unit' (UCU) characterised by single bed acuity flexible rooms where services ranging from admission to electrocardiograms, specimen collection, chest X ray and ultrasound, treatment preparation, step-down recovery, observation and emergency services (overflow) could all be provided. Cells would ebb and flow throughout the UCU based on time of day and patient demand. Patient flow could be managed by a medical 'concierge' to ensure services were flowing as planned.

At this point, teams really began seeing the potential of designing for optimised workflow in meeting several objectives. If workflow is optimised and unnecessary redundancy is removed from the physical process, then less space is needed to resolve the process. If less space is programmed, travel distances can be shortened and cycle times improved. If cycle times are improved, throughput is increased per full-time equivalent (FTE) and so on.

Two additional charrettes (bringing the total number to six) were held in early December and January to examine physical building system alternatives. Some alternatives presented included long span inverse truss structural systems (for the D&T areas of the facility), tilt-up construction for the in-patient tower, HEPA filtered induction units for patient rooms, exterior duct distribution systems (concealed by false fronts), variable refrigerant volume (VRV) systems or modified heat pumps for D&T areas, decentralised mechanical and electrical (M&E) distribution located close to the areas they serve in the extrastitial space created by the inverse trusses, extensive use of LED lighting, light harvesting, cogeneration and use of photovoltaics. Extensive standardisation of components also led to a concept of maximising manufacturing and minimising construction. By standardising virtually all headwalls, one type for in-patient rooms, another for out-patient rooms, these can all be pre-manufactured and then put in place rather than being stick-built in place.

7.4 Results and conclusion

The resulting deliverables, submitted on 31 January 2007 envision dramatic improvements typically exceeding the expectations set at the beginning of this process. All three teams learned from one another. Teams evolved their own collaborative model in the process, each benefiting from the work of the other two teams. This learning was pervasive, benefiting all of the consultants as well as the participants on behalf of Sutter Health as well.

Moving forward, Sutter Health intends to advance many of the building systems alternatives through more study and refinement, continuing a set-based approach to project development until decisions are required (the last responsible moment) to continue to move the project forward. Preliminary evaluation suggests that this project may provide a framework for developing optimisations in workflow which result in 46% improvements overall, including reduced discreet steps in the overall care process, reduced aggregated cycle times for studied outcomes and resource use per APD, 35% less space per APD, 30% overall less construction costs per APD and 25% less energy use for area of the facility.

It is said that necessity is the mother of invention. Never before has the pressure to find innovative lean concepts been greater in both industries: construction and healthcare. As evidenced in the work done for the Sutter Health Prototype Hospital Initiative, the potential of lean thinking in these industries can lead to dramatic transformation in a time when it is so greatly needed.

References

Ward, Allen, Jeffrey K. Liker, John J. Cristiano and Durward K. Sobek II (1995). "The Second Toyota Paradox: How Delaying Decisions Can Make Better Cars Faster". *Sloan Management Review*, Spring, pp. 43–61.

The Strategic Service Development Plan

An Integrated Tool for Planning Built Environment Solutions for Primary Healthcare Services

Ged Devereux

8.1 Introduction

This chapter focuses on the findings of a 2008 comparative case study that examined the effectiveness of the Strategic Service Development Plan (SSDP) to support the development of built-environment solutions for primary care services. The intention of the study was to examine and compare three UK National Health Service (NHS) public sector partners that make up the Manchester, Salford and Trafford (MaST) LIFT Partnership (Manchester Primary Care Trust – PCT, Salford PCT and Trafford PCT) and how they had used the SSDP process to identify good practice in the delivery of the SSDP to support partnership working, the planning process and benefits realisation.

8.2 Background

Delivering healthcare is complex, involving a range of variables including mixed delivery settings, a plurality of commissioners and providers, changing organisational priorities, changing patterns of usage and of course the health sectors history of re-structuring. This requires a number of plans to deliver services in such a complex field of service delivery.

The SSDP is a plan introduced by the Department of Health (DoH) in 2002 to facilitate innovation in service and capital investment in primary care. In April 2002 six local PCTs in partnership with the three local authorities and Greater Manchester Ambulance Service published the first MaST SSDP. The aim of the SSDP was to improve primary care estate and the access to primary care services in MaST. In 2002 the SSDP was seen as an instrumental planning tool to deliver modernised patient-focussed community health and social care services.

Although 90% of patient contact with the NHS is for primary care services and it is usually the publics' first contact with the healthcare system, investment in primary care infrastructure has historically been inadequate and has not been carried out on a systematic basis (NAO, 2005). The bulk of public sector investment has been channelled into secondary and tertiary sector infrastructure. The result has seen a decline in the quality and fitness for purpose of the primary care infrastructure (MaST LIFT, 2002).

The MaST LIFT SSDP (2002) revealed the poor quality of existing primary care estate. Primary care and social care estate was deemed to be non-compatible with developing service models (MaST LIFT, 2002):

- approximately 80% of primary and healthcare premises were below recommended national guidelines;
- less than 5% of general practitioner (GP) services were close to pharmacy or social care provision;
- over half of all GP services were not purpose built, being adapted from retail or residential housing stock;
- few premises complied with the requirements of the Disability Discrimination Act (DDA, 1995).

The MaST sample sites were chosen as the case study sites as these three organisations had been working in partnership to develop individual SSDPs and a MaST wide SSDP over a similar period of time. The aims of the study were:

- to analyse the partnership working arrangements in place to develop and implement the SSDP process in the case study organisations;
- to analyse how the planning process underpins the delivery of the SSDP in the case study organisations;
- to analyse the SSDP planning process to demonstrate and capture benefits for users, providers and commissioners of primary healthcare services.

This study sought to explore the following themes, which are described as follows:

- the development of primary healthcare;
- the role of the built environment in primary healthcare;
- the origins of SSDP.

8.3 The development of primary care

Peckham and Exworthy (2003) suggest that primary care occupies the central ground in the UK NHS and has become a major focus of health policy. They argue that since 1997 health policy has shifted from an emphasis on secondary hospital-based care to primary community-based care. Lewis and Dixon (2005) consider this central role of primary care and suggest that the government's approach to delivering change within the NHS is increasingly reliant on the introduction of market mechanisms and that primary care will play a pivotal role in applying market incentives to other parts of the health system.

The concept of primary healthcare adopted at the World Health Organisation (WHO) sponsored Conference of Alma Ata in 1978 provides a comprehensive working definition of primary healthcare that has been accepted by developed and developing countries. A progressive primary healthcare approach is one that (WHO, 1978):

- challenges the society to address the socio-economic causes of poor health and makes provision for basic health needs;
- encourages community empowerment (ensuring that people are fully able to manage resources that are available to them);
- provides comprehensive quality healthcare including promotive, preventative, curative, rehabilitative and palliative services;
- demands concerned and accountable health worker practice;
- prioritises the people who are most disadvantaged ensuring that healthcare is accessible, equitable and affordable to all;
- recognises the importance of integrated service provision from primary to tertiary levels of care within a coherent health system;
- promotes inter-disciplinary, multi-professional and intersectoral collaborative teamwork for development.

Iliffe (2001) characterises primary care as the industrialisation of family medicine. However other writers argue that primary care is much broader than medical services. In the UK, for example, there exists a long and rich history of community nursing and community medicine provided in partnership with social care and welfare services (Ottewill and Wall, 1990).

Starfield (1998, pp. 8–9) provides a useful definition of primary care as:

That level of health service system that provides entry into the system for all new needs and problems, provides person – focussed (not disease orientated) care over time, provides for all but very uncommon or unusual conditions, and co-ordinates or integrates care provided elsewhere or by others.

However it remains difficult to see how many of the activities provided in primary care are different from those provided in other sectors of healthcare. Starfield (1998) contends that the unique factor may be that primary care is the gatekeeper of NHS services and acts as the single port of entry into the planned NHS with the notable exceptions of accident and emergency and other unplanned care services such as sexual health services. Figure 8.1 illustrates the differentiation between primary and secondary care.

Klein (1990) observes that the NHS resembles an opera with recurring melodies and changing characters; this has some relevance to the changing nature of the primary care field. Peckham and Exworthy (2003) support this view arguing that primary healthcare remains a system in flux suggesting that the health system has co-existed as two distinct elements for many years even prior to the formation of the NHS in 1948. They maintain that a strand of generalist medical services and community health services remained fragmented from each other and the broader health services for the first 20–30 years of the formation of the NHS. Not until the 1990s and through a series of policy developments such as GP fundholding

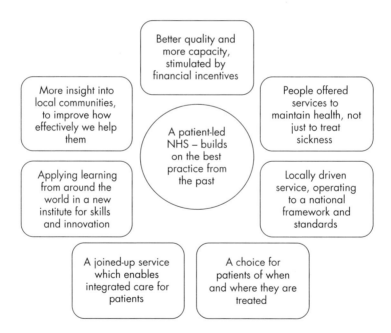

Figure 8.1   Primary and secondary care.

Source: The NHS improvement plan – putting people at the heart of public services. Department of Health (2004). Crown Copyright material is reproduced with the permission of the Controller of HMSO and the Queen's Printer for Scotland.

and the emergence of primary care groups (later to become PCTs) did primary care services become an integrated part of the NHS.

Building on Klein's opera analogy Harrison and Dixon (2005) suggest that primary care has moved 'centre stage' and that the roles of the characters has changed whereby GPs have again become commissioners through practice-based commissioning and other practitioners such as nurses have also changed role. We might even consider the role of the patient to have changed as they are now considered as consumers of healthcare (Lewis and Dixon, 2005). As opposed to passive recipients of health and social care policy has shifted to recognise the informed consumer of health and social care, this policy shift has become known as 'a patient led NHS' (DoH, 2005a), illustrated in Figure 8.2.

Figure 8.2 also highlights the continuous review that the NHS is undergoing. At the heart of this profound process of transformational change is the vision of a service focused around the patient (a patient-led NHS), delivering against national standards and showing clear improvements in the quality of care provided. At the same time the NHS was challenged to adapt to a rapidly changing environment, grasping the opportunities offered by technological and medical advancements and managing pressures, particularly on the workforce (Palmer, 2006).

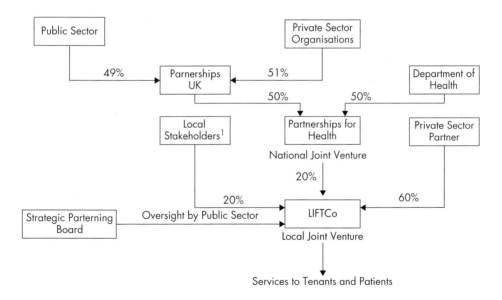

Figure 8.2 A patient-led NHS.

Source: Creating a patient-led NHS – Delivering the NHS improvement plan. Department of Health (2005a). Crown Copyright material is reproduced with the permission of the Controller of HMSO and the Queen's Printer for Scotland.

The emergence of a range of settings such as the community and the home has affected the way that partners involved in the SSDP consider approaching partnership working, integrated planning and how shared benefits have been agreed and measured across the plurality of service providers.

The rapid pace of change within primary healthcare services shows no sign of slowing. This is typified by Lord Darzi's Next Stage Review (Darzi, 2007). There is no doubt that this continual change has had a detrimental effect upon planning within the NHS (Palmer, 2006). However it is also clear that the following key challenges have emerged for the NHS and will therefore be key requirements of emerging SSDPs (Darzi, 2007):

- ensuring clinical decision-making is at the heart of decisions about the design and future of the NHS;
- improving care by increasing quality and making services better integrated;
- making care more convenient, easier to access, delivered in the right place and integrated across primary and secondary care;
- building a service which is based around patient control, choice and local accountability and less on central direction.

8.4 The role of the built environment in delivering primary healthcare

The built environment has a major effect on the way individuals live their lives with regards to physical and mental well-being (CABE, 2002). The modernisation agenda for healthcare services places great emphasis on improving the patient experience; the quality of healthcare buildings has an important part to play in achieving this goal.

In 2006 DoH patient consultation identified a general dissatisfaction with healthcare premises, which must be rectified (DoH, 2006a). The overriding aim is to provide a healing environment. This concept is based on evidence that links the design of the built environment to patient recovery, showing that a better environment contributes to better outcomes (NHS Estates, 2002). At the same time the consultation re-affirmed the notion that a better working environment improves staff morale and contributes to both recruitment and retention of staff (DoH, 2006a).

The MaST LIFT SSDP (2002, p. 15) also highlights the relevance of the built environment:

Service development is often severely hampered by the limitations of the premises available to deliver health care. Investment in primary care facilities in the past has tended to be on a piecemeal basis, not resulting from a strategically identified service need. Rarely has it been designed to achieve integrated service delivery between different health care organisations and social care providers. GPs face significant disincentives to practicing in inner cities, including restrictive long-term leases and where existing premises are often poor and inappropriate for the delivery of modern primary care services.

Delivering the NHS Improvement Plan (DoH, 2004) has challenged the NHS, and some aspects of its implementation have proven difficult given the configuration of the built environment for healthcare. In 2005 the National Audit Office (NAO) put this challenge into context by finding that investment in primary care was inadequate with most public sector finances being channelled into the hospital sector (NAO, 2005). The NHS Improvement Plan announced the governments intention to enter into new forms of partnership to finance the improvement of primary care premises (DoH, 2004). This need for modernisation was coupled with a growing governmental realisation that government was no longer able to finance all the required investments for public service modernisation. The answer seemed to lay in the involvement of the private sector.

The debate over the success and appropriateness of private finance initiatives (PFI) and private and public partnerships (PPP) tends to generate an ideological debate when applied to the NHS. At the centre of this debate is the notion that the NHS is, and should remain a universal service that is free at the point of delivery (Palmer, 2005). These concerns were echoed by the Democratic Health Network (DHN, 2005) when they voiced concern about the possibility of the development of a secondary market with the inclusion of the private sector in the planning and development of primary care facilities. The DHN maintain that this is an issue that continues to cause controversy in relation to PFI/PPP deals in both health and education.

Government policy regarding the introduction of PPP into the built environment for healthcare suggests that private sector partners can provide solutions for weakness in the public sector and can serve three identified purposes (Lewis and Dixon, 2005):

- increasing available funding for primary care built environments;
- widening the range and quality of built environment solutions for primary care;
- strengthening the quality and capacity of management skills to develop built environment solutions for primary care.

As early as the mid-1970s central government looked to introduce the concept of PPP's into the NHS and the provision of health and social care (Doyal *et al.*, 2000). By 2005 the NHS growth in partnership working was established and the NHS was buying more than £1.2bn of care services a year from the private and voluntary sector (Kings Fund, 2001).

As part of the NHS Plan 2000 ambitious targets for investment in primary care premises were announced with the main vehicle for investment being a PPP, this partnership is known as the Local Improvement Finance Trust (LIFT). LIFT was announced in 2001 and was based upon a series of PPP companies that were referred to as LIFTCo's. Each LIFTCo acts as a joint venture made up of local stakeholders (typically PCTs, local authorities and GP's) private sector partners and government agencies. Each LIFTCo takes ownership of the premises it builds or refurbishes and leases the space back to the public sector partner through the PCT (NAO, 2005). The structure of the LIFT PPP sought to extend the practice of PPPs as illustrated in Figure 8.3.

As a strategic partnering tool the NAO (2005) concluded that LIFT was beneficial to private and public sector partners alike. Other procurement methods in the health field such as 3rd party developments provided an ad hoc design and build solution that did not guarantee if developments fitted with the wider strategic direction of NHS and associated public sector services designed to improve the health and well-being of the local population.

Critics of the PPP and its application to health services contest that the debate over their success in primary care has remained un-focussed blaming the government for failing to specify the particular 'problems' for which the private sector is supposed to provide 'solutions' (NHS Alliance, 2001).

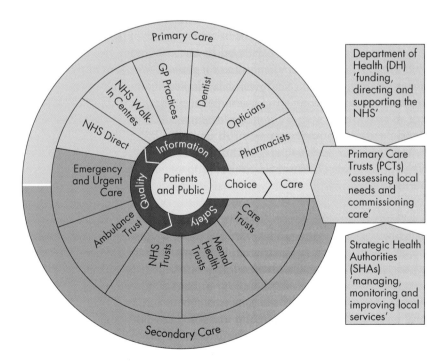

Figure 8.3 Structure of a LIFT PPP.

Source: Innovation in the NHS: LIFTs. Report by the Controllerand Auditor General/HC 28 Sessions 2005–2006. National Audit Office (2005). Crown Copyright: reproduced by permission of the National Audit Office.

Eight years after the introduction of LIFT as a PPP for investment the built environment for primary care it is not clear if LIFT is having the impact expected by the DoH. In 2000 the NHS Plan had expected LIFT to be at the forefront of delivering the following strategic targets for investment in the built environment for primary care (DoH, 2000):

- up to £1bn investment in primary care facilities by 2010;
- up to 3000 family doctors premises substantially refurbished or replaced by 2004;
- five hundred one-stop primary care centres built by 2004.

In fact in 2006 the NAO concluded that LIFT was not close to achieving any of the above targets. Instead the NAO found that the role out of LIFT was slower than expected and this was in part due to a lack of local strategic vision and the relative newness of integrated planning processes to support the delivery of built-environment solutions for primary care. In effect there was little evidence that the SSDP process, which was supposed to provide and articulate local strategic vision and its associated processes to support integrated planning, were being developed and effectively used.

The debate over the effectiveness of LIFT as a PPP for primary care also suffers from poor evaluation of the first 8 years of LIFT delivery. The DHN (2005) criticises the LIFT programme for not having an effective evaluation framework.

The NAO (2005) concluded that developing a strong SSDP process would be crucial in achieving the local and national aims of LIFT programmes.

8.5 The origins of the strategic service development plan

In 2000 the National Primary and Care Trust Development Programme issued initial guidance on the requirement for PCTs to develop SSDPs to underpin and prioritise strategic estate investment. Initial guidance suggested that the SSDP be viewed as a capital investment plan that (National Primary and Care Trust Development Programme, 2000):

- will be a whole system document that reflects service and capital investment requirements across a whole health community (or part of a health community);
- should reflect the plans and aspirations of all users and carers in health and social care facilities, with an emphasis on primary care;
- will demonstrate engagement with local authorities (in particular social services) and the voluntary sector to identify opportunities for better integration of services (not exclusively health services);
- acts as a joint planning document that requires individual stakeholders to commit specific resources to premises development;

- can act as basis for premises procurement, whether as GP- or PCT-led schemes. This could be for a single third party developer scheme, for a LIFT partner, or PCTs batching a number of building proposals together to form the main elements of a memorandum of information that can be issued to all parties who respond to an Official Journal of the European Union (OJEU) notice or other tendering process;
- will describe procurement priorities and proposals for phased investment with target delivery dates in order to reflect future capital/revenue investment requirements to support business cases.

In 2001 the DoH produced its first guidance note setting out the purpose of the SSDP process (NHS LIFT, Starter Pack August, 2001). The SSDP recognised that the introduction of new procurement methods such as LIFT required a whole systems approach to planning. The first version of the Manchester Salford and Trafford SSDP was produced in April 2002 and set out a vision that stated that *'the project board of the local NHS LIFT project is committed to improving both the primary care estate and the access to services in Manchester, Salford and Trafford'* (MaST LIFT, 2002, p. 4).

The NHS Plan (2000) and subsequent guidance notes produced in 2002 by a range of NHS agencies outline the nature of the SSDP process as being (NHS LIFT, Starter Pack August, 2002, p. 20):

> *based on the local HIMP's (Health Improvement plan) and national and local priorities. Detailed audits will be required on Estates, HR, IM&T. Short-term capital investment proposals together with their revenue consequences and service development proposals are required with a demonstration of affordability. Summaries of medium and longer-term proposals are also required. The SSDP is signed up to by all relevant stakeholders and is approved by all relevant agencies.*

The National Audit Office (2005) suggests that the SSDP should set out the ways in which local strategic partnerships (LSPs) of public, voluntary and commercial service providers will supply primary care services having made the most of opportunities to innovate to improve the quality and outcome of services. It follows that if commissioners have successfully articulated how and why they wish to develop health and social care services they will be better placed to match these requirements against the current available facilities to provide services. Therefore the SSDP process and its requirements will impact upon the current and planned built environment for primary healthcare.

In 2006 Partnerships for Health (currently known as Community Health Partnerships) produced pragmatic guidance for preparing an SSDP (Richardson Executive Solutions, 2006). This guidance sets out a whole systems approach for the SSDP to adopt. This guidance recognises that health and social care planning alone will not reduce ill-health and that as a broader strategic planning partnership LSPs are committed to tackling the wider determinants of poor health such as education, employment, crime, poor housing and environment (see http://www.ojec.com). The Partnerships for Health Guidance suggests that the SSDP sets out (Richardson Executive Solutions, 2006):

- new methodological ways to deliver services that are proposed by local users and provider of services that include those services that cut across organisational barriers;
- the practical application of current guidance, initiatives and evidence-based best practice;
- local expertise (consumer, clinical and strategic);
- inclusion of all sectors (public, voluntary, community and private) in the provision of services;
- anticipated and required workforces;
- expectations of built environments that can provide the right environments for new models of care;
- expectations for improved performance of buildings and facilities because of developments in building technology;
- information technology developments and their potential impact on service delivery;
- medical technology advancements and their potential to impact on the nature and location of service delivery;
- financial opportunities and constraints.

8.6 A comparative case study of the MaST LIFT SSDP

The 2008 study of the MaST SSDP's (Devereux, 2008) focussed upon the evaluation of partnership working, the planning process and benefits realisation which have been seen by the NAO as being crucial to the effective development of an integrated approach to healthcare planning (NAO, 2005). The NAO looked at the SSDP process from a national perspective and the DoH adopted Gateway Reviews to look at LIFT and the SSDP from a local MaST LIFT perspective. The following NAO and Gateway findings informed the 2008 study when considering partnership working, the planning process and benefits realisation.

8.6.1 Partnership working

The 2004 Gateway Review identified that the LIFT programme and the SSDP process are based upon the effective delivery of a complex partnering arrangement between public and private sector organisations. The review found that the MaST LIFT Strategic Partnering Board (SPB) lacked the necessary leadership skills to forge these partnerships. Furthermore it concluded that the SSDP was the responsibility of the SPB and as such it is this group of public sector leaders that need to take account for the development of the SSDP process. In a national audit of LIFT programmes of work the NAO (2005) carried out a review examining expenditure on NHS LIFT on behalf of the House of Commons Committee of Public Accounts. The NAO found the effectiveness of the SPB structure in six sample sites across England to be ineffective (NAO, 2005) which supports the findings of the Gateway Review of MaST LIFT (2004).

The 2004 Health Gateway Review found that a weakness of the partnering relationship is that it is sometimes seen as something soft and not a tangible product. Whereas it is true that partnership is about good relationship building and human interactions. The Gateway Review (DoH, 2004) recognises the need to strengthen this area of work by developing explicit systems to support partnering work such as a shared risk log between the public and private sector partners and an agreed negotiations framework. There is a perception that partnership 'happens' however it must be recognised by the SPB that partnership has to be created (DoH, 2004).

8.6.2 Planning process

In 2006 the second Health Gateway Review of MaST LIFT suggested that the SSDP process requires tight and timely project management across all partners. The review concluded, '*It is not clear how effective the current SSDP process is in engaging service users and compiling their views*' (DoH, 2006b). A further conclusion of the report was that it is not clear how the current SSDP engages clinicians and other service providers in the planning process. This is increasingly crucial as the NHS develops a plurality of providers and introduces elements of contestability that will inevitably attract 3rd party providers into the health and social care market. Therefore the SSDP must be a consultative document that engages a wide range of service users and providers (DoH, 2006b). It must also actively engage other sectors of the NHS especially acute trusts. If it is the expectation that secondary hospital services will increasingly be provided in community settings then there seems to be a good case to engage the acute hospital trusts in the development of the SSDP process and the future configuration of health services (DoH, 2006b).

8.6.3 Benefits realisation

The 2006 Gateway Review of MaST LIFT found that the evidence base for the current MaST LIFT SSDP appears to be weak. Section 6 of the MaST LIFT SSDP (2006–2011, V2.00) does set out critical success factors, including a requirement that 'health and service outcomes are clearly defined as part of the business case for schemes and a mechanism for undertaking benefits realisation' (MaST LIFT, 2006, p. 13). However the SSDP does not set out the evidence base for the proposed range of facilities and there is an acknowledgement in Section 12 of the SSDP that the partnership needs to evaluate what it is doing and fully articulate the benefits to all stakeholders, including service users and service providers (DoH, 2006b). Similarly, the House of Commons Committee of Public

Accounts (2005) acknowledges that little work has been done to evaluate the impact of the national LIFT programme. The report suggests that proxy measures related to key local health problems identified in the SSDP could be developed linking service developments as a result of the SSDP to a reduction in poor health linked to specific conditions such as cancer and diabetes.

8.6.4 What was learnt?

The 2008 study (Devereux, 2008) into the effectiveness of MaST SSDP processes used a comparative case study approach to enable the researcher to focus on the circumstances, dynamics and complexity of each SSDP. Document analysis and semi-structured interviews were used as data collection tools.

Document analysis of the three SSDPs allowed a qualitative content analysis to identify common factors across the three documents. The process informed three semi-structured interviews with key individuals from the relevant organisations:

* Manchester PCT – Acting Head of Primary Care
* Salford PCT – Acting Head of Commissioning
* Trafford PCT – Heather, Director of Strategic Commissioning

The following three SSDP documents represented the current thinking on the development of built environment solutions for primary care using the SSDP planning framework across the MaST LIFT footprint.

* Trafford Primary Care Trust, Strategic Service Development Plan, Draft 2 (2007);
* Salford Primary Care Trust. A Strategic Service Development Plan for Services in Salford, Salford Primary Care Trust (2007);
* Manchester Primary Care Trust Strategic Service Development Plan (2008).

8.6.5 Common themes of the document analysis

The following findings list common themes taken from the comparative document analysis of the three sample organisation SSDPs and informed the semi-structured interviews.

8.6.5.1 Partnership working

* The SSDP remains a PCT-driven process with limited input from external partners during the early planning stages of the SSDP process.
* All three SSDP processes identified the need to develop partnership structures to support the development of the SSDP.
* There is an inconsistent approach to the involvement of stakeholders such as clinicians, service users and carers.

- Each of the three sample site SSDPs acknowledged the need to develop shared benefits and common value streams with all stakeholders.

8.6.5.2 Planning process

- Evidence of clear SSDP links to wider PCT planning processes.
- SSDP links to external planning processes was mixed and only evident at a strategic level.
- Lack of internal PCT resources to support the SSDP process, with few dedicated resources to develop and implement the SSDP.
- Evidence of emerging methodologies to prioritise built-environment schemes for primary care services.
- Project management of the planning process was ill-defined.
- Value streams are identified through the SSDP process, however there was no consistent approach across the three sample site SSDPs in identifying value for partners and service users.
- Value is often articulated in strategic terms, however there are few examples of these terms being operationalised through the SSDP process.

8.6.5.3 Benefits realisation

- No clear evaluative framework has been developed as part of any of the three sample site SSDPs.
- Lack of integrated performance protocols means that the measurement of SSDP success is difficult.
- All three sample site SSDPs demonstrated clear concise aims and objectives.
- Value has been largely determined by DoH-driven service delivery targets.
- The three sample site SSDPs did not articulate value from the perspective of the user of primary care services.
- Benefits to partner agencies was articulated but was inconsistent across the three sample site SSDPs.

8.6.6 Common themes from the interviews

The following describes common themes taken from semi-structured interviews with key individuals from the three sample organisation SSDPs.

8.6.6.1 Partnership working

- Respondents presented differing views as to what they believed the purpose of the SSDP was. These views ranged from seeing the SSDP as a partnership document, an estate plan and a tool for engagement.

- Only one respondent felt that the SSDP process had successfully engaged partners.
- All respondents agreed that the SSDP process had failed to effectively engage clinicians and service users. Respondents felt that organisations were nervous about the impact on the organisation of open-ended consultations with members of the public and powerful clinical groups.
- Respondents identified a lack of understanding with regards to engaging partners. All respondents recognised the value in engaging partners but also felt that their respective organisations lacked the skills to engage partners effectively.
- However all three PCT respondents gave details of emerging/existing formalised partnership structures to support the SSDP process.
- None of the respondents felt that long-term shared benefits had been identified through the SSDP process.

8.6.6.2 Planning process

- All respondents felt that a clear planning process existed to support the SSDP.
- However respondents also felt that the planning function of the PCT could be improved by becoming embedded across the PCT and by engaging partners more effectively.
- Each respondent identified dedicated project management capacity to support the SSDP. Although all three agreed that the project management function was isolated in the PCT, focussed on capital works and not fully engaged with all divisions of the organisation.
- The links to external planning processes was seen as weak.
- It was agreed that the SSDP process is weakened when the commissioning intentions of the organisation are unclear.

8.6.6.3 Benefits realisation

- All respondents felt that the SSDP set out core values that indicate the future shape of service provision based upon models of preventative and personalised healthcare.
- None of the respondents felt that an evaluative framework had been sufficiently developed to support the SSDP process.
- Respondents felt that respective SSDPs attempted to identify benefit and in one case provided a methodology to support scheme prioritisation. However all respondents accepted that benefits realisation would be difficult to track without an effective evaluative framework.
- All respondents were able to articulate what benefits they believed would be achievable if the SSDP was implemented.

8.6.7 Discussion

The following discussion draws together the findings taken from the 2008 comparative case study into the effectiveness of the SSDP and its processes to develop built-environment solutions for Primary Healthcare Services (Devereux, 2008).

8.6.7.1 Partnership working

The overall purpose of the SSDP remains confused amongst the sample organisations included in the study and seems to deviate from the guidance set out in 2000 by the National Primary and Care Trust Development Programme. The study found that, at its most basic the SSDP is considered as an estate plan and, at its most developed, it is considered a partnership plan to deliver modernised public services. The inconsistent approach to the SSDP across the close working partnership of MaST LIFT is cause for concern. This finding builds upon Lewis and Dixon's view (2005), which suggests that the pace of change in primary care has occurred at such a pace that there is greater scope for organisational and geographical differences in policy process and policy outcome even across similar NHS organisations.

The ability to develop partnership working arrangements to support the development of the SSDP process has been limited across the three analysed SSDP processes. In 2004 the Gateway Review of MaST LIFT found that the strategic leadership of LIFT and the SSDP process was weak. This study has found that the strategic leadership and partnering to develop the SSDP has improved and that there is clear ownership of the SSDP process from strategic leaders.

However, it is clear that this leadership has not fully transmitted to the operational elements of the SSDP process. The study found that the partnership structures that have been established are immature and require substantial support from the partner agencies if they are to succeed.

This study also confirms the 2004 Gateway Review findings in that the SSDP process does not appear to have developed partnering skills to support the SSDP process. The data gathered suggests that the skill of engaging partners and developing effective partnership working is assumed and appears to be lacking in the SSDP process. The Gateway Review (DoH, 2004) recognised the need to strengthen this area of work by developing explicit systems to strengthen partnering work such as a shared risk log between the public and private sector partners and an agreed negotiations framework.

In analysing the partnership working arrangements of the three sample organisations there is little evidence to suggest that the SSDP processes have sought to fully involve clinicians, patients and carers. This appears to contradict the policy shift that has placed the patient at the centre of a patient-led NHS whereby patients are now considered as consumers of healthcare (Lewis and Dixon, 2005).

From the data gathered it is apparent that the development of long-term shared benefits is a priority but remains a problematic task given the lack of commissioning clarity on behalf of PCT organisations. The National Primary and Care Trust Development Programme (2000) maintain that the SSDP should be a whole system document that reflects service and capital investment requirements across a whole health community. Given the data gathered from the three sample organisations this would appear to be very difficult as in all cases the commissioning strategy was not fully articulated.

8.6.7.2 Planning process

In all three sample organisations it was clear that internal planning processes exist to support the SSDP and that each organisation has identified dedicated project management capacity. This approach was endorsed in 2006 by the second Gateway Review of MaST LIFT, which suggested that the SSDP requires tight and timely project management across all partners (DoH, 2006b). However, the study reveals that individuals involved in the SSDP process and questioned as part of the study believe that the links to external planning processes is weak and that the project management skills that are evident are not embedded throughout the different divisions of the PCT. The project management links commissioning, service provision and estate management.

Again, as referred to under the partnership working theme, the absence of a clearly articulated commissioning strategy that incorporates a clear set of commissioning priorities built upon an agreed model of care and an accepted care pathway appears to be the single biggest barrier to developing the SSDP process. This is a real threat to developing an effective SSDP process and, as Richardson Executive Solutions (2006) suggest, it is essential to establish evidence-based commissioning strategy to inform the SSDP process.

8.6.7.3 Benefits realisation

The SSDP documents that have been analysed all make explicit a core set of values based upon a model of preventative and personalised healthcare. The literature available suggests that the identification and articulation of organisational and shared benefits would add value to the partnership approach of the SSDP (Bradley, 2006). This can lead to the measurement of benefits in terms of cost effectiveness to the organisation such as improvements in the efficiency of the care pathway that a patient is on. Shared organisational benefits may include 'softer' benefits such as patient perception and staff morale (DoH, 2006a).

One aspect of this study has been the consideration of value and how value is expressed through the SSDP process. The 2005 Atkinson Review looked at productivity and value across government departments and concluded that value from the perspective of the service user should be given greater weighting when considering the benefits realisation capture regarding the SSDP (Atkinson, 2005).

The literature suggests that there has been little or no thought given to the evaluative processes of the SSDP as suggested by the House of Commons Committee of Public Accounts (2006). This is supported by the findings of this study. There was no evidence of a developed evaluative framework to support the SSDP process. It follows that the SSDPs reviewed did not fully capture benefits for users, providers and commissioners of primary healthcare services.

A key finding of the 2008 study has been that there is no evidence of a developed evaluative framework to support the SSDP across the three sample organisations. This weakness in the sample organisations SSDPs was originally identified in 2006 by the House of Commons Committee of Public Accounts and is supported by the findings of the 2008 study.

8.7 Conclusion

- There is no evidence of a systematic approach from PCTs in developing the SSDP.
- The SSDP process has limited success in engaging partners.
- The partnership structures are immature and require further support.
- The SSDP processes have not effectively involved clinicians, patients and carers.
- The partnership working skills are assumed and appear to be lacking.
- The development of long -term shared priorities remains a priority.
- The lack of clear commissioning strategy requirements continue to hinder SSDP development.
- All SSDPs include clear core values.
- The clear PCT planning processes and dedicated project management skills exist to support the SSDP process.
- The project management needs to be supported across the entire PCT.
- The SSDP benefits are articulated in service delivery terms and fail to articulate benefit from the perspective of the service user.
- There is no evidence of a developed evaluative framework to support the SSDP process.

The study referred to in this chapter has informed local planning for future SSDPs and has led to the adoption of the following recommendations.

8.8 Recommendations

Recommendation 1 Partnership Working – All those involved in the provision of health and social care and related services should be consistently involved in the preparation and development of the SSDP through the formal processes of the Strategic Partnership Board. Public Sector Chief Executive officers should be required to fully participate in the work of the SPB and this must involve more active engagement with the private sector partners. The SPB should commission a piece of work to identify the roles and responsibilities of senior stakeholders and should identify the support that SPB members require to become informed commissioners.

Recommendation 2 Planning Process – The SSDP process should be treated as a project in its own right. As such an appropriate senior responsible owner should be appointed in each partner agency to act as the lead officer responsible for that organisations contribution to the SSDP thereby forming the SSDP project team. The SSDP project team should have appropriate delegated authority from the SPB and this should be connected to the wider planning process.

Recommendation 3 Partnership Working – As part of the SSDP process organisations should promote the full participation of those involved in the provision of services and those that receive services in the early preparation of the SSDP. The SSDP may wish to engage with formal representative bodies such as clinical groups and patient engagement structures as well as wider consultation on a regular basis. The SPB should also consider how these views are represented at the SPB level.

Recommendation 4 Planning Process – The relationship between the SSDP, the local development plan and the commissioning plan of an organisation needs to be made explicit, whereby the SSDP becomes part of the formal planning cycle.

Recommendation 5 Planning Process – The SSDP is intended to be a whole systems planning tool, as such it should be integrated into the planning cycle of the broader local area partnership becoming an adopted planning tool of the local area agreement.

Recommendation 6 Benefits Realisation – In the preparation of the SSDP partners should be brought together in service modelling workshops to devise individual service sector development plans and delivery plans. The proposed SSDP project team should lead this process where a number of service areas are reviewed and remodelled. This will start to re-focus the SSDP away from a facilities focus to a service development focus that may or may not have implications for the built healthcare environment.

References

Atkinson, G. (2005). *The Atkinson Review*. National Statistics Office, HMSO, London.

Bradley, G. (2006). *A Practical Guide to Achieving Benefits Through Change*. Gower Publishing Ltd., Aldershot, London.

CABE. (2002). *Commission for Architecture and the Built Environment. Buildings and Spaces, Why Design Matters*. CABE, London.

Darzi, A. (2007). *NHS Next Stage Review Interim Report*. Crown Copyright, London.

Democratic Health Network. (2005). LIFT: What you Need to Know and What you Need to Ask: Briefing for Non-Experts. London Local Information Unit, London.

Department of Health. (2000). *The NHS Plan – A Plan for Investment, a Plan for Reform*. HMSO, London.

Department of Health. (2002). *NHS LIFT Starter Pack Guidance Note: Department of Health*. HMSO, London.

Department of Health. (2004). *MaST LIFT Gateway Programme Review – Strategic Assessment 8 November – 12 November 2004*. HMSO, London.

Department of Health. (2005a). *Creating A Patient Led NHS – Delivering the NHS Improvement Plan*. HMSO, London.

Department of Health. (2006a). *Our Health, Our Care, Our Say: A New Direction for Community Services*. HMSO, London.

Department of Health. (2006b). *MaST LIFT Gateway Programme Review – Strategic Assessment 28 March – 31 March 2006*. HMSO, London.

Devereux, G. (2008). *A Comparative Case Study into the Effectiveness of the Strategic Service Development Plan (SSDP) and the Processes to Develop Built Environment Solutions for Primary Health Care Services*. Research Institute for the Built and Human Environment School of Construction and Property Management, University of Salford, Salford, UK.

Disability Discrimination Act. (1995). *Office of Public Sector Information*. HMSO, London.

Doyal, Y., Bull, A., and Keen, J. (2000). Role of private sector in United Kingdom Health Care System. *British Medical Journal*, V321, 563.

Harrison, R. and Dixon, J. (2005). *Regulation in the New NHS Market*. Kings Fund, London.

House of Commons Committee of Public Accounts (2005), 'http://www.publications.parliament.uk/pa/cm200506/cmselect/cmpubacc/uc634-iuc63402.htm, House of Commons, London.

House of Commons Committee of Public Accounts. (2006). *NHS Local Improvement Finance Trusts. House of Commons Committee of Public Accounts*. The Stationary Office, London.

Iliffe, S. (2001). The National Plan for Britain's National Health Service. Towards a Managed Market. *International Journal of Health Services Research and Policy*, V31(1), 105–110.

Kings Fund. (2001). *Public Private Partnerships and Primary Care*. Kings Fund, London.

Klein, R. (1990). The State and the Profession: The Politics of the Double Bed. *British Medical Journal*, V301, 700–702.

Lewis, R. and Dixon, J. (2005). The Future of Primary Care. Kings Fund, London.

Manchester Primary Care Trust. (2008). Strategic Service Development Plan. Manchester Primary Care Trust, London.

Manchester, Salford and Trafford LIFT. (2002). *A Strategic Service Development Plan for Health Services*. MaST LIFT, Salford.

Manchester, Salford and Trafford LIFT. (2006). *Strategic Service Development Plan # 2*. MaST LIFT, Salford.

NHS LIFT, Starter Pack (2001) Starter Pack, Department of Health, Crown publications.

NHS LIFT, Starter Pack (2002) Starter Pack, Department of Health, Crown publications.

National Audit Office. (2005). *Innovation in the NHS: Local Improvement Finance Trusts. A Report by the Controller and Auditor General. HC. Session 28 – 2005–2006*. HMSO, London.

National Health Service Estates. (2002). *The Design Brief Working Group: Advice*. NHS Estates. Citigate Lloyd Northover, London.

Ottewill, R. and Wall, A. (1990). *The Growth and Development of Community Health Services*. Sunderland Business Education Publishers, Sunderland.

Palmer, K. (2006). NHS Reform Getting Back on Track. Kings Fund, London.

Peckham, S. and Exworthy, M. (2003). *Primary Care in the UK: Policy Organisation and Management*. Palgrave Macmillan, Basingstoke.

Richardson Executive Solutions Ltd. (2006). *Pragmatic Guidance for Preparing Strategic Service Development Plans: Partnerships for Health*. Richardson Executive Solutions Ltd., London.

Salford Primary Care Trust. (2007). *A Strategic Service Development Plan for Services in Salford*. Salford Primary Care Trust, Salford.

Starfield, B. (1998). *Primary Care: Balancing Health Needs Services and Technology*. Oxford University Press, New York.

Trafford Primary Care Trust. (2007). *Strategic Service Development Plan Draft 2*. Trafford Primary Care Trust, Trafford.

World Health Organisation. (1978). *Declaration of Alma Ata: International Conference on Primary Health Care, Alma Ata, USSR*. World Health Organisation, Geneva.

Part 2

Academic Contributions

From Care Closer to Home to Care in the Home

The Potential Impact of Telecare on the Healthcare Built Environment

James Barlow, Steffen Bayer, Richard Curry, Jane Hendy and Laurie McMahon

9.1 Introduction

Many billions of pounds are being spent modernising the UK's healthcare infrastructure. Investment programmes for new primary care facilities, and new acute and community hospitals are underway (NHS Executive, 1999). When completed, the previously outdated built environment for healthcare will largely be replaced.

Parallel to this investment, a major overhaul of the way health services are provided is underway. In the coming decade, a decentralisation of services, including acute and diagnostic work, from in-patient hospitals and into community settings – not only closer to people's homes but to the home itself – seems inevitable (Department of Health, 2006).

What impact might this have on future infrastructure requirements? Discussion on this question is increasingly urgent, given that the choices that are being made today will determine the built infrastructure for healthcare for decades to come. This chapter outlines the key trends that are likely to impact on the future demand for healthcare built infrastructure, focusing on the introduction of 'telecare'.

9.2 Key trends

The interaction of healthcare services, technologies and built infrastructures is complex. Changes in one element can lead to unpredictable or highly lagged effects on the others. First, it is necessary to consider the key trends – a mix of demographic, social, policy and technological factors – which are likely to influence the future healthcare built environment.

Increasingly people wish to have access to a wider range of higher quality healthcare services closer to where they live and work – they are less willing to travel to where it suits the system to provide them. This is reinforced by the consistent policy statements on taking services out of larger hospitals and delivering them in the community, including existing community hospitals. And the right of patients to choose which hospital to be treated in may result in more dynamic and competitive hospital trusts decentralising services both to retain existing patients and to act as 'collecting points' in others' catchment areas. Decentralising services to satellite settings also has the potential advantage of helping to reduce fixed costs in expensive acute hospitals, while at the same time maintaining a revenue flow.

Implementation of 'practice-based commissioning' of health services by general practitioners (GPs)[1] may also reinforce the trend for the formation of large-scale 'GP collaboratives'. These will not only have the facility to commission services that are provided by local hospitals but also potentially the scale to provide them closer to the patient through small community hospitals and health centres.

Technology is also playing an important part in decentralising services. The introduction of progressively less invasive clinical treatments and clinically administered diagnostic procedures means that doctors need not necessarily work in settings where there is access to multi-specialist teams and high-technology equipment (Ofcom, 2008). Developments in facilities technology is making it increasingly safe and acceptable for diagnostic tests and acute clinical treatments to be delivered in settings beyond large hospitals – even in mobile and modular facilities. The digitalisation of both pathology and imaging mean that patients need not travel to hospitals for the majority of their diagnostic tests. The tests can be administered locally and the results can be read and interpreted anywhere. We are already seeing the impact of this technology with the decentralisation of access to diagnostic services and the centralisation of the processing, analysis and interpretation.

In this chapter we focus on the mainstream implementation of 'telecare' – care that allows people to be cared for remotely in their own homes. This particular innovation has gained high levels of policy interest because of its potential ability

---

[1] For details see: http://www.dh.gov.uk/en/Managingyourorganisation/Commissioning/Practice-basedcommissioning/index.htm

to shift services from traditional institutional settings, raising questions about the design and role of future healthcare infrastructure.

9.3 What is telecare?

Faced by rising demand and capacity constraints, governments and healthcare providers in many countries are turning towards information and communications technology (ICT) to help support and enhance existing services. Moves towards ICT-enabled care have also been stimulated by innovation in various underpinning technologies – sensors, information processing, user interfaces – and by the falling costs and rising availability of fixed and mobile telecommunications.

Telecare is a form of ICT-assisted care provision which potentially reduces the risks associated with care provision outside formal institutional settings and allows improved data collection on an individual's changing care requirements (Barlow *et al.*, 2007a). Terms such as 'telecare', 'telemedicine' and 'telehealthcare' are often used interchangeably. We define telecare as the use of ICT to provide health and social care directly to the end-user – a patient or someone in need of care and support. This excludes ICT-enabled applications designed for exchanging information solely between professionals, generally for diagnosis or referral – we define this as 'telemedicine'.

Some forms of telecare involve the use of the Internet or telephone to provide better *information and support*, for example to patients in a particular locality and with a particular health condition. Another type of telecare focuses on *monitoring* an individual's vital signs or their activities of daily living through the use of sensors. This allows a response to an immediate need such as a fall or a sudden change in an individual's vital signs to be triggered or longer-term evidence of a change in their condition to be gathered. In this way the risks associated with care outside formal care institutions can be better managed. Telecare systems should be flexible and expandable to meet changing individual needs (Audit Commission, 2004).

Telecare is based on the premise that people in need of care should be able to participate in the community for as long as possible, supporting them at home, in 'lower intensity' residential care settings or on a mobile basis in the normal daily living environment as appropriate. Because it can potentially transform a previously unsuitable environment into a lower risk one, telecare has important implications for the future location of care delivery.

In the UK calls for telecare have been made in numerous government and other official documents since the late 1990s (Barlow *et al.*, 2005), and over £170m has been made available during 2006–2009 to support new telecare services. Telecare is increasingly embedded in key health policies relating to managing chronic disease better and providing people with greater choice over their care pathways, as well as targets aimed at reducing the number of inappropriate admissions and facilitating earlier discharge.

9.4 The impact of telecare on care services

As yet, the impact of telecare on health and social care services is unclear. It has been shown to have potential in ameliorating many of the factors, such as family or caregiver stress or a fear of falling, which trigger increased levels of care and support (Brownsell *et al.*, 2005). And although the quality of the evidence varies considerably, there are numerous scientific studies which report clinical and quality of life benefits for conditions such as diabetes, hypertension and congestive heart failure (e.g. Amala *et al.*, 2003; Cappuccio 2004; Klemm *et al.*, 2003). A recent systematic literature review found almost 9000 papers (as at January 2006) in scientific journals reporting on outcomes of telecare trials (Barlow *et al.*, 2007b; CSIP, 2006; cf. Bensink *et al.*, 2006; Paré *et al.*, 2007).

The various systematic reviews demonstrate that evidence base for the impact of telecare *across the care system* – on patient flows or use of resources – is far less robust. However, some studies suggest there are possible cost savings arising from reduced hospital admissions for certain health conditions. A major problem is that most reported telecare trials are of small-scale pilot projects and as such are limited in the extent to which they can provide reliable indications of the systemic impact of mainstream telecare services. Furthermore, because telecare is essentially a hybrid technology and service innovation aimed at supporting people in the wider community, its introduction spans health, social care and housing authorities and the distribution of cost and benefits between these authorities may be uneven.

Simulation modelling of the processes involved in delivering care to different populations or patient groups can help to identify the effect mainstream telecare might have (e.g. Bayer *et al.*, 2007; Cox *et al.*, 2008). Such projections are liable to a great degree of uncertainty as they can only be as good as the data and assumptions on which they are based, but they do allow us to ask questions about how care services might need to evolve as telecare is introduced. For example, modelling based on data from telecare schemes to increase safety and security in the homes of frail older people, and assuming telecare is only introduced to new patients suggests that the number entering residential care across the UK as a whole could be reduced by 5% five years after the introduction of telecare. The effect of telecare over 5 years in this scenario is relatively limited because of the time required for substantial numbers of this cohort of elderly people to become so frail as to require institutional care. In the longer term the effect of telecare would be far more pronounced – perhaps 20% over 10 years and 25% over 20 years. Monthly hospital admissions for this population could drop over 5 years by around 3% and then stabilise (Bayer *et al.*, 2007).

The true impact of telecare on care services will only begin to emerge as it is rolled out and schemes are fully evaluated. So far, progress has been slow in all countries but there is now momentum behind its introduction – some 17% of US home care agencies are thought to use some form of remote monitoring system (NAHCH, 2008) and in the UK 17% of the population over 65 has remote monitoring (OC&C, 2005). It therefore seems likely that it will play an increasingly significant part in shifting the balance of care towards the community and people's own homes.

9.5 Implications for the healthcare built infrastructure

If telecare is seen as a key part of the future care system, local bodies responsible for planning and delivering services should ensure that care pathways are redesigned not only to anticipate future needs but to adopt telecare as part of those pathways. The introduction of telecare should in itself become a focus for better integration of health and social care services. With highly dependent populations, such as frail elderly people or those with complex long-term chronic conditions, there will be a need for integration with a wider range of care organisations than for those with relatively simple care needs.

From the outset telecare should also be designed into proposals for modernising or replacing healthcare built infrastructure. This process will need to embrace primary and secondary care, as well as social and housing services, to ensure new investment is suited to local needs and affordable for the local care economy as a whole.

Much effort is expended on addressing the internal functions and design of acute hospitals to meet changing healthcare practices and needs. Individual clinical services within proposed new hospitals are mapped against care process and design solutions which minimise travel times, encourage efficient use of staff and respond to new ways of working are developed. However, these solutions will increasingly need to consider telecare as this opens up possible new models for care delivery and for its supporting built infrastructure.

The government proposals to invest in 'community hospitals' – smaller scale and more widely distributed than large district general hospitals – represent an opportunity to explore and introduce new care and infrastructure models which embrace telecare. Planning new community hospitals and reconfiguring existing ones will therefore require primary care trusts to consider models of care – and associated infrastructure – that best suit their local needs.

So what might the future community hospital look like? There is no standard model for a community hospital in terms of their role within local care systems, funding sources and age and design characteristics, nor is government proposing to be prescriptive in the new investment programme. There is therefore no 'one size fits all' model, but we can nevertheless identify several broad areas where telecare could be combined with community hospital services:

- There is a clear role for the community hospital as a facility for local, more accessible health services such as blood and urine testing, imaging and minor injuries treatment. The first four facilities to receive funding in the UK (Washington Primary Care Centre in Sunderland, Gosport War Memorial

Hospital, Yate Health and Children's Integrated Services Centre and Minehead Community Hospital) are projected to perform around 25 000 medical tests, treat 30 000 minor injuries and handle 20 000 out-patient appointments every year (Building, 2006). However, some of this workload will inevitably decline as technologies for near patient testing grow in sophistication – home blood sugar monitors are now a reality – and become integrated into telecare services.

- There is scope for day case procedures to be delivered by community hospitals. These need to be coupled with responsive, flexible intermediate care and community rehabilitation services, with telecare providing better support after discharge by monitoring the patient's recovery and providing an 'electronic safety blanket' for those without adequate support networks at home.

- The community hospital could represent an important node in the bundle of services designed to enable older people to remain in their own homes. Around 60% of admissions to residential and nursing homes are direct from hospital. With recuperation and rehabilitation many older people could regain enough of their mobility and daily living skills to return home. There is therefore a role for the community hospital as an appropriate rehabilitation facility where older people's long-term care needs can be properly assessed and appropriate packages of telecare introduced. Experience from existing telecare trials has shown that helping patients and carers understand and get used to telecare is a key part of successful adoption and there may be scope for community hospitals to set up 'telecare demonstration suites' for this purpose.

How will new government health and social care policies, changing public expectations and technological innovation impact on other parts of the built environment? One important area might be in the modification of people's housing requirements and by challenging preconceived notions of what type of housing provision is appropriate for people's varying needs. In particular, the sheltered housing and residential and nursing care sectors – which house between 10% and 15% of elderly, disabled and vulnerable people – may need to change their role. This is because of the way telecare potentially creates new care pathways and changes the number of people proceeding along existing pathways (Bayer *et al.*, 2007). For example, there is a demand for residential home placements following hospitalisation and from people who have developed conditions that cannot be adequately managed in the community. By including telecare as part of a care package, this demand can be slowed or reduced and people diverted to either their own home or sheltered accommodation. They might then stay at home for the rest of their life or until they need hospitalisation. In this way telecare reduces the need for relocation as people's care needs change or delays their move into some form of institutional care setting (Barlow and Venables, 2004).

Telecare may also have an impact on the type of people living within the sheltered housing and residential and nursing care sectors. We argued above how simulation modelling has suggested that the need for places in this sector may fall considerably. An alternative picture would be that the overall numbers living in this sector remains stable or only declines slightly, but the characteristics of the population changes. Even though they are living in a sheltered community the effect of telecare for residents is to maximise their level of independence and reduce the worry that they may have to leave should their capabilities decrease too far. This is turn may result in different types of resident – perhaps with higher levels of need – being accommodated in residential and sheltered/extra care housing than is currently the case, providing primary care trusts and social services with an additional care placement option.

9.6 Conclusion

The shift of services out of large acute hospitals is likely to accelerate as telecare becomes more common. This presents an opportunity to rethink how best to use scarce resources more effectively and at the same time create more responsive and accessible services for all. However, it also raises some important questions about the future built infrastructure for healthcare.

Hospitals, in particular, lie directly on the fault line arising from the reconfiguration of services from acute to community settings and the rationalisation of emergency and specialist facilities.

We know that we will no longer need nearly as many large-scale district general hospitals, but what will replace them? The government's commitment to invest in a new generation of community hospitals recognises that advances in technology could radically shift thinking about how care can be provided. But what do hospitals of the future do and look like? Are clinical adjacencies still going to be as important? And what kind of built environment will be required in community settings?

It is clear that the mainstream housing stock will become an increasingly significant part of the care delivery system, but this needs to be 'fit for purpose'. While telecare potentially facilitates the delivery of far more personalised care packages to people in their preferred domestic environment, it may well increase the need for measures to improve the physical quality of the mainstream housing stock such as improvements to disabled access or thermal efficiency in order to make it suitable for frail people – no amount of telecare can compensate for an inadequate, poor quality home. This poses important questions about the cost of upgrading and adapting the housing stock and whose budget this comes out of.

Bringing care closer to home – and ultimately within the home itself – is a reality. This is the result of important policy changes and the introduction of enabling technologies and associated service innovations such as telecare. This illustrates the dynamics of the interaction between changes in healthcare technologies, services and infrastructures. It also highlights how the current investment in acute hospital infrastructure might prove to be inappropriate in the future.

9.7 Acknowledgements

This work was funded by the Engineering and Physical Science Research Council and partly conducted through the Health and Care Infrastructure Research and Innovation Centre.

References

Amala L, Turner T, Gretton M, Baksh A, Cleland J (2003) A systematic review of telemonitoring for the management of heart failure. *The European Journal of Heart Failure*, 5: 583–590.

Audit Commission (2004) *Implementing telecare. Strategic analysis and guidelines for policy makers, commissioners and providers.* London, Audit Commission.

Barlow J, Bayer S, Curry R (2005) Flexible homes, flexible care, inflexible attitudes? The role of telecare in supporting independence. *Housing Studies*, 20: 441–456.

Barlow J, Bayer S, Curry R (2007a) Assessing the impact of a care innovation: telecare. *System Dynamics Review*, 23: 61–80.

Barlow J, Singh D, Bayer S, Curry R (2007b) A systematic review of the benefits of home telecare for frail elderly people and those with long-term conditions. *Journal of Telemedicine and Telecare*, 13: 172–179.

Barlow J, Venables T (2004) Will technological innovation create the true lifetime home? *Housing Studies*, 19: 795–810.

Bayer S, Barlow J, Curry R (2007) Assessing the impact of a care innovation: Telecare. *System Dynamics Review*, 23(1): 61–80.

Bensink M, Hailey D, Wootton R (2006) A systematic review of successes and failures in home telehealth. *Journal of Telemedicine and Telecare*, 12(S3): 8–16.

Building (2006) *£44.5m in funding released for community facilities.* 21 December 2006.

Brownsell S, Aldred H, Hawley M (2005) *The role of technology in addressing the care and support needs of older people.* 5th International Conference of the International Society for Gerontechnology, Nagoya, Japan, 24–27 May 2005.

Cappuccio F, Kerry S, Forbes L, Donald A (2004) Blood pressure control by home monitoring: Meta-analysis of randomised trials. *British Medical Journal*, doi:10.1136/bmj.38121.684410.AE (11 June 2004).

Cox B, Barlow J, Bayer S, Petsoulas C (2008) *Modelling service innovation in stroke care.* Final report to the Department of Health (Information and Communication Research Initiative 2), October 2008.

CSIP (2006) *Building an evidence base for successful telecare implementation – Updated report of the Evidence Working Group of the Telecare Policy Collaborative chaired by James Barlow – November 2006.* London, Care Services Improvement Partnership.

Department of Health (2006) *Our care, our say: A new direction for community services. (Command Paper 6737).* London, Department of Health.

Klemm P, Bunnell D, Cullen M, Soneji R, Gibbons P, Holecek A (2003) Online cancer support groups: A review of the research literature. *CIN: Computers, Informatics, Nursing*, 21: 136–142.

NAHCH (2008) *National study on the future of technology and telehealth in home care.* National Association for Home Care & Hospice, Philips Home Healthcare Solutions, and Fazzi Associates. Available at: http://www.philips.com/HomeCareStudy

NHS Executive (1999) *Public private partnerships in the National Health Service: The private finance initiative.* London, Treasury Public Enquiry Unit.

OC&C (2005) *Initial findings on international telecare.* OC&C Strategy Consultants, London.

Ofcom (2008) *Health technology scenarios and implications for spectrum. Health socio-economic study; Technology scenarios development – Annexes.* London, Office of Communications.

Paré G, Jaana M, Sicotte C (2007) Systematic review of home telemonitoring for chronic diseases: The evidence base. *Journal of the American Medical Informatics Association*, 14: 269–277.

Risk Management and Procurement

Nigel Smith, Denise Bower and Bernard Aritua

10.1 Introduction

Historically in the UK during the first half of the 20th century engineering contracts were largely derivatives of property contracts and healthcare was local and fragmented. With the post-war changes the National Health Service (NHS), was established and the *Institution of Civil Engineers* (ICE), produced their first model Conditions of Contract which were designed to be used in a consultant-led design and contractor-led construction procurement system. For many years this was considered a reliable strategy for the industry although requiring a long duration. As a result of efforts to adopt more integrative approaches to procurement, the main government agencies launched new and more collaborative procurement systems such as prime contracting, design and build and the variants of the private finance initiative (PFI) that enabled integrated delivery of design and construction through a single point of contact acting as the leader of a supply chain.

NHS ProCure 21 is the Health Service's response to the challenges set down in Sir John Egan's report *Rethinking Construction* (1998) and HM Treasury's *Achieving Excellence* document (OGC, 2000). The key purpose of the ProCure 21 approach was to enable the NHS to achieve best client status, establish a partnering programme through long-term framework agreements, promote the use of high-quality designs and ensure continuous improvement (DoH, 2000). The government commissioned a task force (*headed by Sir Malcolm Bates*) to undertake tow reviews of public sector procurement as part of the wider agenda to improve value for money (VFM). The two *Bates* commissions (HM Treasury Taskforce, 1999) set the agenda for the ProCure 21 partnerships which were intended for publicly funded schemes with a value of over £1m and all PFI schemes between £1m and £20m. NHS ProCure 21 involves NHS clients working in partnership with substantially fewer suppliers. To do so the NHS establishes framework agreements with carefully selected companies (principal supply chain partners, PSCPs) from within the construction industry. Through the collaborative arrangements the PSCPs develop and consolidate long-term relationships with the NHS as they bid for work through the framework agreements on the basis of best practice and VFM rather than least bidding price.

This chapter provides a comprehensive overview of the management of risk through the NHS procurement process. It examines the use of collaborative procurement and the multi-project approach to delivery of projects and concludes with views of the future for sustainable procurement practice.

10.2 General principles of risk management in infrastructure procurement

Much has been written about the nature and terminology of risk and uncertainty, therefore only a brief overview will be presented here. The word 'risk' is derived from the Latin 'riscare' meaning 'to dare' which promotes views of opportunities, success and failure. In construction-related activities, the definitions of risk in common usage may be divided into three categories:

- The common usage of the term 'risk' in dictionaries and the definitions provided by a number of professional bodies and standards tend to view risk as wholly negative. Standards such as the *Norwegian Standard NS 5814* (1981), *British Standard BS 8444-3* (1996) and the *National Standard of Canada CAN/CSA-Q850-97* (1997) have references to adverse consequences and negative effects as key features of risk. Although some definitions such as that provided by the *Institute of Risk Management* (2002) do not explicitly use the terms in the aforementioned standards, phrases such as the 'chance of bad consequences' or 'exposure to mischance' lend themselves to this school of thought.
- The UK *Association for Project Management* (APM) in its *Project Risk Analysis and Management* (PRAM) *Guide* (2004) adopts a more neutral and broad definition for risk without specifying whether the consequences are positive or negative. The *British Standard BS 6079-3* (2000), the Japanese standard for risk management *JIS Q2001* (2001) and the joint *Australian/New Zealand risk management standard AS/NZS 4360* (2004) share this general view which could encompass both upside and downside effects of risk.
- Other guidelines explicitly introduce the concept of including upside effects in the definition of risk. For example the *Risk Analysis and Management for Projects* (RAMP) *Guide* (2005) views risk as a threat or opportunity which could affect achievement of objectives either adversely or favourably. Similarly

the *Guide to Project Management Body of Knowledge* (PMBoK) (2004) produced by the *Project Management Institute* (PMI) and the *APM Body of Knowledge* (APMBoK) (2006) define risk with recognition of the possibility of a positive or negative effect. The *ISO/IEC Guide 73:2002 – Risk Management Vocabulary: Guidelines for use in Standards* (2002) defi1nes risk as a combination of a probability of an event and its consequence; noting that the consequences may be positive or negative.

Nevertheless, Chapman and Ward (2003) argue that a definition that does not include both downside and upside consequences of risk focuses teams on reducing underperformance at the expense of losing out on potential benefits. This is interesting and highlights the need for a broader view of risk. To manage risk requires a methodology, as it is not the elimination of risk, or the prediction of the future, nor is it a prescriptive process. Risk management is the creative process of making better decisions for tomorrow, or at least decisions we regret least tomorrow. The chapter adopts this view.

Classification of risk is an important part of the decision-making process since it forms the basis for identification, assessment and response to risks. Many approaches have been suggested in the literature for classifying risks in construction-related projects. No standard model for classification will guarantee identification of all the risks facing a project or organisation, but adopting some sort of classification will help produce better, more complete and more consistent sources of information on which to base decisions. For example, Smith (2003) proposes the categorisation of risk into *global* and *elemental* risks. In this approach, global risks refer to the more general risks which might influence a project but may be outside the control of project parties whereas elemental risks refer to risks which are associated with elements of the project, namely implementation and operation risks; and for some projects there will be financial and revenue risks. These risks are more likely to be controllable or manageable by the project parties.

In the UK general guidance on the risk management process is provided by codes and methodologies produced by public and learned society organisations. For example, HM Treasury have a guide on risk: *The Orange Book* (2004), the ICE produced the *RAMP Guide* (2005), the APM has produced its *PRAM Guide* (2004). Fortunately all these documents rely upon the same basic principles which Figure 10.1 captures as a generic risk management cycle. As shown in Figure 10.1 the process of risk management should be continuous and iterative allowing for *planning, identification, assessment, response and monitoring and control.* Such a formal approach allows the project team to respond to specific risks. In addition, it provides a mechanism for analysis and management of risks and also enables better understanding of the project and common language in communicating among participants (Simister, 1994). This, in turn, leads to more realistic planning and decision making.

Figure 10.1   The risk management cycle.

10.2.1 Risk planning

The planning phase is the step that exhibits widest variability in terms of scope and level of detail. At one extreme there are standards that take a broad view and include in this step organisation wide issues such as establishing the risk management policy, defining role and responsibilities at various levels and establishing the processes to be followed (Chapman and Ward, 2003; Fairley, 1994; Harrington and Niehaus, 1999; Head and Horn, 1997; Kliem and Ludin, 1997). At the other end there are those that follow a more focused approach, consisting simply of planning the application of an existing risk management process to a specific project or instance. With exceptions the difference seems to be whether the process of managing risk is approached from a strategic top-down perspective or a project level down-top perspective.

10.2.2 Risk identification

It is generally accepted that risk identification determines which risks are likely to affect the project and documents the characteristics of each one. Risk identification should address both the internal and external risks. The main methods of risk identification in construction-related projects are well known and include: brainstorming; use of cause and effect diagrams (Ishikawa/fishbone diagrams); checklists; concept and cognitive mapping; Delphi technique; decision trees; and influence diagrams (Raz and Micheal, 2001).

10.2.3 Risk assessment

Risk assessment is a means to systematically use the information obtained from the identification process to characterise and analyse the risks, and make a judgement about the likely consequences. Risk assessment may be classified as qualitative or quantitative (Smith and Min, 2006; Smith *et al.*, 2006).

Qualitative risk analysis is the first stage in any risk assessment process and one of the most useful stages of the management process since it lays the foundation for all the subsequent stages. According to Smith *et al.* (2006) a typical qualitative risk assessment usually includes the following issues:

- a brief description of the risk;
- the stages of the project when it may occur;
- the elements of the project that could be affected;
- that factors that influence it to occur;
- the relationship with other risks;
- the likelihood of it occurring;
- how it could affect the project.

Frequently, no further analysis needs to be done; however some elementary assessment may be carried out to assist the decision-making process often to provide some kind of ranking, sorting, classifying and preliminary quantification. This may be in the form of probability-impact grids, sensitivity analysis, decision trees or influence diagrams (Flanagan and Norman, 1993; Institution of Civil Engineers and the Actuarial Profession, 2005; IRM/ALARM/AIRMIC, 2002; Smith *et al.*, 2006).

Quantitative risk assessment may be in form of probability *theory-based analysis* which attempts to specify a probability distribution for each risk, then considers the effects of the risks on economic parameters in combination (Vose, 2000).

Two terms that are often used in probability analysis are modelling and simulation. *Modelling* is the process of describing the project in a mathematical way. A model can never be a perfect representation of reality as the actual reaction of the project to certain factors is unknown (Rowe, 1994). However, it is important that the model is as realistic as possible to ensure that it will react in a similar way to the project when real-life complications are added. *Simulation* involves testing and experimentation with models and measuring the effects (Raftery, 1994). Monte-Carlo or Latin Hypercube sampling methods is often used to reflect the randomness in risk models (Vose, 2000). Risk analysis software such as *@risk*, *predict!* and *crystal ball* have been designed to assist in the quantitative analysis process.

A major challenge in using quantitative risk assessment methods is that the majority of tools and techniques are descriptive and qualitative in nature, and there are few risk identification tools that are based on statistical or mathematical techniques. This is reasonable since risk identification is a stage that involves project stakeholders; most of whom would not be adept at using advanced mathematical modelling. Furthermore, some risks are intangible and are therefore difficult to quantify. As a result the risk analyst must find a way to express these qualitative outputs of the risk identification process into quantitative terms that may be modelled using mathematical approaches; and express the outputs in terms that will be helpful in aiding decision making. Additionally, risk identification is often undertaken at the early stages of the project when information is limited, imprecise/vague or even substantially non-existent. The application of artificial intelligence methods such as *fuzzy reasoning* and *artificial neural networks* may be used to represent risk through membership functions by characterising vague and subjective situations (Ross, 2004). Approaches such as *info-gap* decision modelling represent more recent and innovative attempts to quantify risk without the need to assign some sort of normalised mathematical function (a probability density function or membership function). Info-gap modelling seeks to quantify the disparity between what the decision maker knows and what could be known (Ben-Haim, 2006).

In brief, however the benefits of carrying out some form of quantitative risk analysis using mathematical methods are indisputable and many organisations that use them confirm their benefits in enhancing better decision making.

10.2.4 Risk response

Risks response is the decision-making phase which indicates the action required to reduce, eliminate or avert the potential impact on a project or business. The common response approaches include avoidance, transference, reduction and provision of contingency (Smith *et al.*, 2006).

Risk management is not a separate, one-off activity. It is fundamental to the successful management of an infrastructure procurement project and is the responsibility of the project management team. The major risks associated with the concession need to be identified, appraised and allocated in an equitable manner following risk management best practice.

10.3 Risk and procurement routes

Procurement is the term used to describe the overarching process of the identification, selection and acquisition of civil engineering services and materials, their transport, their execution or implementation and subsequent project performance (Bower, 2003). Procurement includes the 'internal' aspects of administration, management, financing of and repayment for these activities. Risk is allocated through procurement routes and contract strategies that assign work and the associated payment mechanism for delivery of infrastructure. Masterman (2002) argues that the essence of procurement decisions is to identify and allocate risks; and all the procurement routes with their associated contract documentation are essentially efforts to manage risk.

Under the traditional procurement systems in construction the process is sequential with the designer being the fist point of contact for the client and often assuming design risks, the contractor takes on construction risks (which may be cascaded to subcontractors through appropriate mechanisms) and another party taking on risks for maintenance and operation of the infrastructure. The management forms of contract represented an attempt at integrating design and construction.

As a result of the *Accelerating Change* (SFC, 2002) and *Modernising Construction* (NAO, 2001) studies, central government departments were advised to secure infrastructure and carry out refurbishment/maintenance projects using three primary procurement routes: private finance initiative (with its variants); design and build and prime contracting. These collaborative routes were conceived as ideal for stemming adversarial relationships that characterised the construction industry and fostering best practice and VFM through early involvement of all participants in the delivery of the infrastructure. The collaborative procurement routes also enabled equitable allocation of risks to the party best able to handle them.

In contractor-led design and build, prime contracting and turnkey procurement approaches contractors get involved much earlier in the design phase when the client's requirements are translated into technical drawings and specifications. Under these procurement systems designers are essentially considered part of the contractor's supply chain. The actual technical design is performed by designers who may be part of the contractor's organisation or as is usually the case, procured by the contractor in some form of service contract. Whatever the case, as the single point of contact for the client the contractor takes the main design and construction risks.

The concept of using private finance for public infrastructure is not new and has been utilised on many well-known projects around the world. There are records of water supply contracts in France dating back to the 16th century and projects including the Suez Canal and the Trans-Siberian-Railway were funded using private finance. During the 20th century the principle of the state raising revenue through taxation and funding public infrastructure became the norm. However in the 1980s after the quadrupling of oil prices and the increasing pressures on public spending the use of private finance was proposed again.

When the PFI was first introduced to the UK construction industry in 1992, it had to be made clear that this was not merely an alternative source of funding but a means to access commercial and entrepreneurial skills of the private sector; as not many public sector bodies had either the expertise or the enthusiasm to use this newly initiated strategy in their building or infrastructure projects (HM Treasury, 2000, 2003). In the PFI approach to procurement the emphasis is not restricted to realising a physical asset but involves the overall success of delivering a service from financing to operating, maintaining and managing. PFI differs from privatisation in that the public sector retains a substantial role in PFI projects, either as the main purchaser of services or as an essential enabler of the project. It differs from contracting out in that the private sector provides the capital asset as well as the services (Allen, 2001). As such, the risk assessment process addresses the investment rather than the engineering (Smith *et al.*, 2006). There are two main objectives behind the PFI: sharing financial risk between the public and private sector and bringing commercial reality to large complex projects. The potential risks of financing, procuring, operating and maintaining are largely passed over to the private sector for the duration of the concession period (Merna and Njiru, 2002).

Typically the private sector promoter (also known as a special project vehicle, SPV) will be a consortium formed specially for the project, usually with minimal asset value. PFI concessions use non-recourse financing which means that the true assets are the cash flows from the project and there is no recourse to SPV assets or parent company assets. The SPV uses secondary contracts to transfer risks to other parties for design, construction, financing operation and maintenance (Merna and Smith, 1996).

From a private sector point of view, investment is motivated by the business opportunity. This means that there must be a minimum return on invested private capital for a particular investment to be attractive and worth the risks (Yescombe, 2002). Conversely, the public sector seeks to balance socio-economic costs and benefits associated with investing in infrastructure. The trade-off between the two is the premise for public–private partnerships (PPP). A key element of a PPP is that the social cost benefit of the project will be positive but the financial analysis may not be rigorous enough for 100% private financing. Ideally a project would be financially attractive and offer positive social cost benefits. However this is not often the case. Some projects with potentially positive social cost benefits are not necessarily attractive to a private sector investor; and yet the private sector may be best suited to manage the risks and deliver the scheme. In these cases the public sector will have to close the bankability gap by retaining risk, providing equity, offering indirect assistance and lastly only in extreme case investing debt finance.

10.4 Risk in NHS procurement

Most of the government departments in the UK public sector, for instance the Department of Health (DoH) and the Ministry of Defence (MoD) (http://www.dh.gov.uk, 8/2/2005 and http://www.mod.uk, 23/3/2005) will only allow their projects to be privately financed if and only if the proposed solution has been justified of providing better VFM than a publicly funded option. Also the proposed solution needs to be within the annual budget allocated for the specific project.

As stated in the NHS PFI Guideline (http://www.healthplan.org.uk (a), 15/3/2005), the client's design variation should be borne by the NHS Trust. The Office of Government Commerce (OGC) (http://www.ogc.gov.uk (a), 15/3/2005) indicates that all risks related to the construction cost overrun are with the private sector. Therefore when some of the risks have been transferred to the private sector, the public sector has to take less *net present value* (NPV) of risk compared to the publicly funded option. In some cases, the *NPV* of risk to the Trust is even being reduced to zero, for instance the land purchase risk, by transferring the risks to the private sector.

All proposed private-financed projects within the NHS require a justification that the PFI would offer better VFM than a publicly funded option prior to receiving an approval from the DoH. This is the premise for the *public sector comparator* (PSC) which estimates the hypothetical risk-adjusted cost if a project were to be financed, owned and implemented by government. *PSC* is meant to meet the need to make a compelling argument in favour of using private finance rather than traditional public procurement.

10.5 Multi-project procurement

Many hospitals may be delivered as single projects but inevitably some have to be procured as multi-projects. Depending on the size and complexity of the projects it may usually be possible to cascade the scheme of delivering projects as part of a multi-project scheme clustering projects with similar objectives, time scales, geographical locations or complementary resources. A multi-project delivery approach offers advantages to both the public and private sector. The public sector benefits from the accelerated investment, an integrated solution and a focused approach to overall objectives of healthcare. On the other hand the private sector would be more willing to invest in schemes with greater critical mass, as such schemes should in principle bring greater scope to offer innovation and deliver more cost-effective solutions in terms of finance, capital, life cycle and operational costs; thereby assisting in risk distribution. Bid costs for both sides reduce as the number of sites increase.

The multi-project environment includes programmes and portfolios which provide the link between the overall strategic objectives and the individual projects (Morris and Jamieson, 2006). In this environment project portfolio management is predominantly about 'choosing the right projects' whereas programme management is about 'doing the projects right' (Blichfeldt and Eskerod, 2008; Blismass *et al.*, 2004; Winter *et al.*, 2006). This means that at the portfolio level the main concern will be the centralised management of one or more programmes and includes identifying, evaluating, prioritising, authorising, managing and controlling programmes, and other related work, to achieve specific strategic objectives (PMI, 2006). Programme management on the other hand will focus on coordinating all activities in the individual projects required to achieve a set of major benefits that would otherwise not be realised if the projects were managed independently (OGC, 2007). To a large extent project portfolio management is mostly strategic while programme management is more tactical. Nevertheless it must be admitted that in practice it may not always be possible to adopt a two-tier multi-project management strategy with distinct roles and indeed the portfolio and programme management roles are often combined.

Significantly, the use of framework agreements permits long-term alliances to be formed that often engender cross-project learning, partnering, continuous improvement and performance benchmarking. Procurement strategies may therefore be adopted for integrating design, construction and asset management teams, and also for maintaining a differentiated contract structure if so required.

10.6 Sustainable NHS procurement options

After the initial growth in PPP projects in the health sector following the two Bates commissions, the real strengths and weaknesses of this procurement strategy are again under review. Despite some successes in delivering viable PPP projects for major health infrastructure and services it is clear that PPP is not panacea for all projects.

There are clear tensions within long duration PPP concessions between VFM, transfer of risk and loss of flexibility. Technology changes, views on treatments change and public expectations change, yet it is unlikely for the payment tariff to be sophisticated enough to cater equitably for all these variations. The ideas of two-tier multi-project approaches outlined in the preceding section may provide a vehicle for a sustainable procurement strategy in future. In particular the 'bundling' together of a range of projects to cover the full spectrum of health services and related infrastructure, but with greater understanding of management options and the types of tariff requirements than operates at present, offers exciting prospects.

In conclusion, it seems increasingly likely that in view of the trends in government policy, the NHS will continue to move from concentrating on treating acute illness and injury to maintaining and enhancing health. This raises interesting issues regarding alternative funding sources currently excluded from procurement options including healthcare insurance funds, life assurance funding and income from the leisure sector possibly including cross subsidies from spas and hotel chains. What is more predictable in the UK is that there is unlikely to be a sudden increase in the amount of public sector finances to investment in health services procurement in the short to medium term. Therefore, it seems that some form of PPP procurement will continue to be used but the mechanisms must continue to evolve to satisfy the long-term requirement for sustainable procurement. As shown in this chapter, risk management is intrinsic to the process of procuring infrastructure. Essentially, the different procurement routes represent different approaches to allocating and managing risk. The reforms in public sector procurement which have resulted in more collaborative and integrated ways of working across the supply chain mean that many schemes are increasing delivered in multi-project environments as a means to obtain benefits that would not be possible if the projects were managed singularly. These developments are the subject of a research agenda to explore new frontiers in multi-project management.

References

Allen, G. (2001) *The Private Finance Initiative.* London, House of Commons Library – Economic Policy and Statistics Section.

APM. (2004) *Project Risk Analysis & Management (PRAM) Guide.* High Wycombe, Association for Project Management.

APM. (2006) *APM Body of Knowledge* (5th edn). UK, Association for Project Management.

AS/NZS 4360:2004. (2004) *Risk Management.* Homebush, NSW, Australia.

Ben-Haim, Y. (2006) *Info-Gap Decision Theory: Decisions Under Severe Uncertainty.* Amsterdam, Elsevier.

Blichfeldt, B. S. and Eskerod, P. (2008) Project portfolio management – There's more to it than what management enacts. *International Journal of Project Management,* 26(4): 357–365.

Blismass, N., Sher, W. D, Thorpe, A. and Baldwin, A. N. (2004) Factors influencing delivery within construction clients' multi-project environments. *Engineering, Construction and Architectural Management,* 11, 113–125.

Bower, D. (2003) *Management of Procurement.* London, Thomas Telford.

BS 6079-3:2000. (2000) *Project Management – Part 3: Guide to Management of Business-Related Project Risk.* London, British Standards Institute.

BS 8444-3:1996. (1996) *Risk Management. Guide to Risk Analysis of Technological Systems.* London, British Standards Institute.

CAN/CSA-Q850-97. (1997) *Risk Management: Guideline for Decision Makers.* Canada, Canadian Standards Association (CSA).

Chapman, C. B. and Ward, S. (2003) *Project Risk Management: Processes, Techniques and Insights.* Chichester, Wiley.

DoH. (2000) The NHS ProCure 21 Initiative.

Egan, J. (1998) *Rethinking Construction (The Report of the Construction Task Force to the Deputy Prime Minister, on the Scope for Improving the Quality and Efficiency of UK Construction).* London, Department of Trade and Industry.

Fairley, R. (1994) Risk management for software projects. *IEEE Software,* 9, 57–67.

Flanagan, R. and Norman, G. (1993) *Risk Management and Construction.* Oxford, Blackwell Science.

Harrington, S. E. and Niehaus, G. R. (1999) *Risk Management and Insurance.* Boston, MA, Irwin McGraw-Hill.

Head, G. L. and Horn, S. (1997) *Essentials of Risk Management.* Malvern, PA, Insurance Institute of America.

HM Treasury. (2000) *Value for Money Drivers in PFI.* London, HM Treasury, Treasury Taskforce (TTF).

HM Treasury. (2003) *PFI: Meeting the Investment Challenge.* London, HM Stationery Office.

HM Treasury. (ed.). (2004) *The Orange Book Management of Risk – Principles and Concepts.* London, HM Stationery Office.

HM Treasury Taskforce. (ed.). (1999) *Review of the Private Finance Initiative by Sir Malcolm Bates: Summary of Recommendations.* London, Stationery Office.

Institution of Civil Engineers and the Actuarial Profession. (2005) *RAMP: Risk Analysis and Management for Projects.* London, Thomas Telford.

IRM/ALARM/AIRMIC. (2002) *Risk Management Standard.* London, Institute of Risk Management, National Forum for Risk in the Public Sector, Association of Insurance and Risk Managers.

ISO/IEC Guide 73:2002. (2002) *Risk Management – Vocabulary – Guidelines for Use in Standards.* International Organisation for Standardization.

Kliem, R. L and Ludin, I. S. (1997) *Reducing Project Risk.* Aldershot, Gower Publishing.

Masterman, J. W. E. (2002) *An introduction to Building Procurement Systems.* London, Spon.

Merna, A. and Smith, N. J. (1996) *Guide to the Preparation and Evaluation of Build Own Operate Transfer Project Tenders.* Hong Kong, Asia Law & Practice.

Merna, T. and Njiru, C. (2002) *Financing Infrastructure Projects*. London, Thomas Telford.

Morris, P. W. G. and Jamieson, A. (2006) Linking corporate strategy to project strategy via portfolio and program management. *International Journal of Project Management*, 25, 57–65.

NAO. (2001) *Modernising Construction*. London, National Audit Office.

NS 5814. (1981) *Requirements for Risk Analyses*. Norway, Norsk Standard.

OGC. (2000) *Achieving Excellence*. In HM Treasury (ed.). London, HM Treasury.

OGC. (2007) *Managing Successful Programmes*. London, The Stationery Office.

PMI. (ed.) (2004) *A Guide to the Project Management Body of Knowledge (PMBOK® Guide)*. Newton Square, PA, Project Management Institute.

PMI. (2006) *The Standard for Portfolio Management. Global Standard*. Newton Square, PA, Project Management Institute.

Q2001:2001, J. (2001) *Guidelines for Development and Implementation of Risk Management Systems*. Tokyo, Japanese Standards Association (JSA).

Raftery, J. (1994) *Risk Analysis in Project Management*. London; New York, E & FN Spon.

Raz, T. and Micheal, E. (2001) Use and benefits of tools for project risk management. *International Journal of Project Management*, 19, 9–17.

Ross, T. (2004) *Fuzzy Logic for Engineering Applications* (2nd edn). London, John Wiley & Sons.

Rowe, W. D. (1994) Understanding uncertainty. *Risk Analysis*, 14, 743–750.

SFC. (2002) *Accelerating Change*. London, Strategic Forum for Construction.

Simister, S. (1994) Usage and benefits of project risk analysis and management. *International Journal of Project Management*, 12, 5–8.

Smith, N. J. (2003) *Appraisal, Risk and Uncertainty*. London, Thomas Telford.

Smith, N. J., Merna, T. and Jobling, P. (2006) *Managing Risk in Construction Projects*. Malden, MA, Blackwell Publishing.

Smith, N. J. and Min, A. (2006) Risk Analysis for Construction Tomorrow. *20th IPMA World Congress on Project Management*. Shanghai, China Machine Press.

Vose, D. (2000) *Risk Analysis: A Quantitative Guide*. Chichester; New York, Wiley.

Winter, M., Smith, C., Morris, P. and Cicmil, S. (2006) Directions for future research in project management: The main findings of a UK government-funded research network. *International Journal of Project Management*, 24, 638–649.

Yescombe, E. (2002) *Principles of Project Finance*. San Diego, CA; London, Academic.

Supporting Evidence-Based Design

Ricardo Codinhoto, Bronwyn Platten, Patricia Tzortzopoulos and Mike Kagioglou

The design of healthcare environments is complex and challenging. Issues to be addressed include the number and variety of users (patients, visitors and staff); the constantly emerging technological changes to support diagnosis, treatment and support services; the high costs involved in providing such services and the nature of the service, which is to care for people's health and well-being when they are most vulnerable. In healthcare there is no room for mistakes. Architects, engineers and healthcare planners do their best in creating design solutions that mitigate the negative impacts that the facility might have upon its users. Several solutions about infrastructure design have been developed around the world through the years and the amount of information currently available is vast, but what does really work?

Aiming to answer that question, evidence-based design (EBD) emerged in the 1980s, as an approach to support designers adapted from evidence-based medicine (EBM). Sackett (1997) defines EBM as the conscientious, explicit and judicious use of current best evidence in making decisions about the care of individual patients. In other words, the aim of EBM is to apply evidence gained from scientific research to support the decision about the most effective and efficient treatment. Therefore, following the same principle, EBD is an approach to assist designers to make decisions about design solutions based on the available knowledge about the impact of those solutions upon people, costs, management, amongst other factors (Malkin, 2008).

But why should we use evidence? The generation and use of evidence has been the focus of many academic debates related to knowledge generation. Issues on the reliability of conclusions based on the use evidence are clear, as highlighted by Johnson and Duberley (2006) in the text of Bertrand Russell:

One day a turkey was hatched. By chance, it stumbled upon corn and water. It was a happy turkey. The next day it happened again and again the next. Being an intelligent turkey, it considered the possibility that supplies might stop and wondered whether it would be necessary to take precautions. It decide to investigate the world to see whether, given a large number of cases and a wide variety of conditions, there were grounds to suppose that the pattern of events so far witnessed would continue in the future. The benefit would be that no precaution against non-supply of corn and water need be taken. After months of careful observations and noting that differences in weather, configurations of the stars, beings encountered, mood and many other things did not stop the supplies, the turkey concluded that the world truly was a wonderful place. The very next day, everything changed. It was 24 December (Russell, 1948).

As argued by Russell (1948) and Johnson ad Duberley (2006), the collection of evidence is far from determining a guaranteed and reliable answer to our daily problems. However, scientific evidence drawn from quantitative and qualitative research seems likely closer to the truth than anything else available. Conversely, one can argue that the use of evidence is not the only step necessary to the design of better services and products. Gathering evidence can be the easiest part in comparison with creating new designs and identifying the worthy problems to tackle. However, the use of an evidence-based approach to design plays an important role within design in mitigating risks related to design decisions.

Moreover, design for healthcare demands a rigorous and complex process involving hundreds of decisions. The employment of EBD for all decisions would be considerably challenging, given the timescales and the number and variety of decisions required. Therefore, evidence should be used to support critical decisions. In those cases, whatever decision is made, post-occupancy studies can be carried out to validate the decision made and subsequently contribute to the evidence-base.

There are several other issues related to the use of EBD. In this chapter the discussions are focused on gathering and compiling evidence about the relationships between healthcare environments and health outcomes. In Section 11.1 the definitions of built environment, healthcare environment and health outcomes are presented. Section 11.2 presents theoretical and practical issues involved in applying EBD. In Section 11.3 the process of searching for evidence is presented. Section 11.4 presents a framework linking the built environment and health outcomes. Section 11.5 presents different ways of compiling and reporting research findings. Finally, in Section 11.6 conclusions are presented.

11.1 Definitions

The built environment is considered to include the surroundings or conditions designed and built through human intervention, where a person, animal or plant lives or operates. The built environment refers to the boundaries that define the 'envelope' of built spaces as well as the inside and adjacent spaces generated and connected to those boundaries.

In this regards, healthcare environments are defined as any specialised building or space where healthcare is delivered and its surroundings such as (but not restricted to) hospitals, primary care centres, hospices and nursing homes.

Finally health outcomes refer to positive and negative measures of physical, mental and social health and well-being such as levels of depression, anxiety and health-related quality of life (HRQL).

11.2 The built environment and health outcomes: considerations about evidence-based design

Before embracing EBD, users must be aware that there are several issues regarding the quality of the existing evidence. The most significant concern arises from the intrinsic epistemological limitation in the field: research methods are simply limited in gathering knowledge from such a complex and dynamic phenomenon because of several reasons, including:

- The lack of explicit cause and effect relationships: this regards the impact that the built environment might have on human experience. In general, existing knowledge has been established by considering different studies simultaneously and examining correlational relationships amongst them. However, this approach can be limited as social, cultural, spiritual and a vast range of other potential variables can affect health outcomes;
- The different strength levels of the evidence: the quality or strength of the evidence may vary given the nature of research methods employed to measure outcomes, ranging from anecdotal to scientifically gathered evidence. Since cause–effect relationships are so complex, scientific evidence is limited in the accuracy of the research findings. Moreover, the descriptions provided in many academic publications is not complete. Generally speaking, details about the population investigated and the environment where the study took place as well as the relevance of the identified results have been, in many cases, omitted;
- The multidisciplinary characteristic of the field: healthcare environments and health outcomes can be investigated from myriad perspectives, for instance: from the psychological point of view related to how people consciously and unconsciously interpret and cope with the surroundings; from the biological point of view considering how the human body reacts to different characteristics of the environment; and from the social point of view investigating how groups and individuals interact within specific settings;
- The fragmentation and sparseness of the knowledge base: because the field is multidisciplinary, there is a vast range of sources of information available. Abundance in this case may be counter-productive as it impacts on transparency of knowledge and challenges the potential to compare results provided by different sources;

- The lack of theoretical consensus: several contrasting theories explain the same phenomenon with no evidence of shared consensus (Codinhoto *et al.*, 2008; Sundstrom *et al.*, 1996). Additionally, those theories may be generated with an interpretivism bias, and therefore reflect individual interpretations that cannot be proven;
- The multitude of measurement methods: there is no commonly agreed methodology for measuring health outcomes resulting from the interaction between humans and the built environment. As a result, it is challenging to integrate existing research findings;
- The number of design variations: changing only one characteristic of the built environment is enough to get a new design configuration. Consequently, too many design variations can exist and to test them all could be an impossible task. In fact, the relationship between the built environment and humans is a 'many-to-many' type of relationship. In other words, the built environment is constituted of several variables (e.g. colours, types of light, spatial dimensions, textures, shapes, artworks) which may have varied and contrasting impacts (e.g. depression, anxiety, relaxation, well-being) in different people (e.g. staff, patients with different illnesses and conditions, varying age, gender, cultural background and embodiment);
- The translation of scientific results into practice: last but not least is the issue related to the translation of academic results into practitioner's language. The application of research findings in practice is a separately defined area of research and there are several models that explain the processes of knowledge generation and its use in practice (Denis and Lomas, 2003; Landry *et al.*, 2001).

Additionally, given that EBD is a relatively recent approach, there are only a few descriptive evidence-bases available (e.g. NHS, 2005; Ulrich *et al.*, 2008), which means that only limited evidence of the specific relationships involving the built environment are available. In response, practitioners are mostly likely to need to develop their own evidence-base until an evidence-base, such as the Cochrane Collaboration, is set. To make an evidence-base strong and reliable, certain steps must be followed. In the Section 11.3 we present recommendations for searching the knowledge base and develop an evidence base.

11.3 Searching for evidence

In EBD the search for evidence can be supported by the use of systematic literature reviews. A systematic review is a review of the literature that follows predetermined rigorous steps that gives strength and *replicability* to the searching process. The full application of systematic review in design is limited, however, is still of a great value.

The process of searching for evidence through systematic reviews can be time-consuming and largely determined by the level of rigorous research applied to the enquiry and whether or not evidence is available. However, following a systematic approach brings *trackability*, allowing the process to be replicated, therefore bringing strength and reliability to the search (Mulrow, 1994; Tranfield *et al.*, 2003).

The most important and relevant step when conducting a systematic literature review is to define a focused research question. Also of importance is determining a template that explicates the variables to be examined along with the establishment of the research question and correlating the findings. The more variables are included in the template, the more likelihood of finding relevant outcomes to the enquiry. Table 11.1 provides an example of a template for an enquiry about healthcare environments and health outcomes. Knowing the parameters in each column and row may help the researcher to define the focus of the research in great level of detail.

Integrative frameworks that map cause and effect relationships to understand the range of issues involved in the research can also be developed to support the searching process. Figure 11.1a, b shows two simplified frameworks. Figure 11.1a shows different built environment variables impacting on

Table 11.1 Generic template for data collection.

Patient group				
Mean Age	Gender	Illness	Treatment	Sample size
Healthcare environment				
Building type	Unit	Setting	Characteristic	
Methodological characteristics				
Research design	Patient blinded?	Study duration	Control group?	
Health outcome				
Psychological	Physiological	Physical	Baseline	

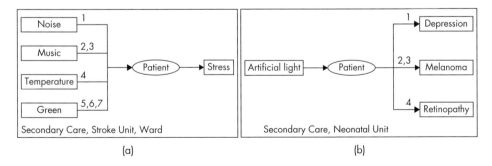

(a)   (b)

Figure 11.1 (a) Built environment variables impacting on stress; (b) artificial light impacting on several health outcomes.

the levels of stress of patients in a stroke unit setting. Figure 11.1b shows the links between artificial light and the development of depression, melanoma and retinopathy. The numbers listed represent the source of evidence and can be listed separately.

Although a narrow research question gives direction to the search process, there is no guarantee of determining relevant findings. Sought evidence, of value for researchers and practitioners might not exist or may not been reported in a similar format or context. In those cases, to consider alternative ways of looking for the same problem can be useful. For instance, changing slightly the focus of the research question or looking for a similar enquiry in a non-healthcare environment.

Once the research question is defined, the second step of conducting a systematic literature review is the definition of the criteria for selecting publications. This can include: (a) the type and form of publication (e.g. peer reviewed journal – ideally – and abstracts), avoiding duplications and including papers published in different languages; (b) a specific period of time, for example within the last 5 years; (c) the research method, for instance, by considering or restricting the research to a specific method used to measure outcomes. The same protocol developed to narrow down the research question can be used as a checklist of the inclusion and exclusion criteria.

The next step of the review is to select the database or databases to search. There are several databases available, some more specialised than others. In general, the search engines of the databases are very similar, however it is recommended to identify potential variables before starting the search as Boolean operators, truncations and wildcards may vary. Usually the use of keywords is necessary and the template may provide the necessary ones. The set of selected keywords used will work as an electronic inclusion and exclusion criteria of publications. Still, several papers might be captured in this first search, and reading through the resulting abstracts and selecting the relevant ones with a basis on the inclusion and exclusion criteria, in general, is enough.

Finally, with the relevant pieces of evidence in hand, the results from different studies can be tabulated, compared and analysed. Synthesising results can vary between narrative reviews and meta-analysis. In medical research, meta-analysis has been used to aggregate the research results (Mulrow, 1994) whereas in management research interpretive and inductive approaches have been used instead (Tranfield *et al.*, 2003).

Regardless of whether reviewers are synthesising direct or indirect evidence, many factors can modify etiologic and prognostic associations, diagnostic accuracy and therapeutic effectiveness. This impact on findings is because of the fact that study participants are often drawn from various settings and have a wide spectrum of baseline risk, disease severity and socio-demographic and cultural characteristics. Therefore, special attention is needed to those details of the study (Mulrow and Cook, 1997). In the next section, a more detailed look into the issues related to organising the findings and ideas that can support the decision making in design is presented.

11.4 Healthcare environments and impacts on health

This section presents frameworks for data compilation developed in a 2-year research project focused on gathering evidence about the impacts of the built environment on health outcomes.[1] In order to achieve the research objectives, a literature review was carried out that was both multidisciplinary in focus — including the social, physical and health sciences — and diverse in scale – that is from overall functioning of hospitals to specific equipment such as air conditioners. Evidence from quantitative and qualitative studies was gathered through an extensive review of academic research about the physical environment and its direct and indirect impacts on health outcomes. The review adopted some aspects of systematic literature reviews, recommended by Mulrow (1994), Boaz *et al.* (2002) and Tranfield *et al.* (2003) as presented in the following paragraphs.

The review started with the identification of previous state-of-the-art reviews including: Devlin and Arneill (2003), Ulrich *et al.* (2004) and NHS Estates (2005). Jointly, these three publications refer to 293 academic journals and provided an overview of the knowledge domains investigating this phenomenon.

The second step was to identify and select databases and keywords to support the search for existing research results. Seven databases (ASSIA, CINAHL, DAAI, OCCL, HMIC, MEDLINE and 'The Safer Environment' Database – NHS Estates, 2005) and 50 keywords were used. In total, 1163 abstracts were evaluated against predefined quality parameters, including the provision of appropriate background, research method, findings applicability and generalisation. As a result, 92 papers were included in this study.

Subsequently, a conceptual taxonomy for data collection was developed to map cause–effect relationships between built environment variables and health outcomes. The framework is presented in no particular order, and by no means it is implied that there are well-defined boundaries in the taxonomy used (e.g. music can comfortably be classified under arts as well as acoustics). As presented in Codinhoto *et al.* (2009) this includes:

- Ergonomics: Ergonomic studies are related to the investigation of how the mind responds and the human body mechanically reacts when interacting with the built environment. It includes not only dimensions and shapes of objects and spaces but also the way these are distributed and organised. Built environment characteristics investigated in ergonomics include, for instance, layout, shape, size and weight of objects and their surface materials and textures and colour;

---

[1] For further information about this project see Codinhoto *et al.* (2008).

- Fabric and ambient: The fabric relates to the 'envelope' setting boundaries around us and includes the design and the materials used in the construction of buildings, the spaces between buildings (e.g. parks) from urban to site scale. For instance, in relation to an object, one can be interested in psychological impacts influenced by textures (e.g. smooth, rough, silky), patterns (e.g. stripes, dots, chequer, plain), colours (e.g. yellow, orange, red, black, white) (Cooper *et al.*, 2008) and dimensions (e.g. size, height, width, depth) as well as the characteristics of included art and design features (e.g. symmetry, balance rhythm, creativity, invention). On the other hand, the ambient relates to the space in which we are immersed and includes acoustics, light and air quality, temperature, ventilation, humidity, views. For instance, in relation to a hospital ward, one can interrogate about the impacts of noise and music, natural and artificial light, high and low pressure and humidity and so for (Cooper *et al.*, 2008);
- Art and aesthetics: In healthcare this area is known in practice as the Arts in Health. The arts are increasingly being utilised in healthcare settings in the recognition that not all aspects of health can be addressed by medicine alone and of the need to improve and humanise the healthcare environment (Lord Hunt, 2002). The nature of the art experience includes a myriad of approaches such as music, live or recorded; drawing; painting and traditional and contemporary art. Reported effects vary from biomedical impacts to psycho-social benefits such as improving self esteem and reducing isolation (Staricoff, 2004). The delivery of the activity and the approach employed will also support different health outcomes. For instance, Art Therapy is a recognised clinical treatment for varying conditions. Therapists are trained to deliver a therapeutic intervention utilising the arts such as drama, painting or music will result in differing effects to that of participatory arts, which focuses on the value of creativity for its own sake and utilises professional artists in a healthcare setting to enable patients to develop creative skills (Smith, 2003). Finally, there are distinctions between a patient's passive experience of artworks installed in a healthcare setting and active participation in an arts activity such as painting which will also impact in differing ways upon health and well-being (Staricoff, 2004);
- Services: Refers to non-clinical operations and continuous maintenance of characteristics and performance of the built environment. This category is also very relevant as the lack of appropriate equipment or maintenance of subsystems may have extreme consequences. For example, the maintenance of ventilation systems, water supply and cleanliness are included in this category.

The framework described in the preceding paragraphs provides insights on how complex the built environment can be and the myriad variables related to it. Structuring information about the built environment can be challenging and time consuming without a pre-understanding of its elements, features and characteristics. In the following section we present examples of how to organise information.

11.5 Organising information

The collection of evidence is an important and insightful step towards EBD. The compilation of evidence may result in fragmented and heterogeneous pieces of information. In the design process this heterogeneity can be explored in different ways depending on the type of project and phase of design development. In this section, we present four different ways of compiling information and discuss how they can be useful to support EBD. These frameworks are not exhaustive and are presented with the aim of generating insights for professionals who want to engage in developing an evidence-base to support their decisions. The first three frameworks are recommended when dealing with specific problems whereas the last one uses a different mapping technique and is recommended for creating an overall picture of existing evidence.

11.5.1 Framework 1: patient groups framework

The objective of organising data according to varying needs of patient groups is to consider all the issues which apply to a specialist unit and their inpatients. Through this approach, particular issues may emerge that relate to the ways in which patients with specific (limited) conditions interact with the surrounding environment. One of the positive aspects of using this approach is that scientific studies may report on specific design solutions (e.g. for tackling disorientation and anxiety of patients with various forms of dementia) rather than generic ones (e.g. guidelines to improve wayfinding). Those solutions will be based on patient experience of the environment and the ideas can be used in any phase of the design development.

Table 11.2 presents an example of data organised according to patients groups. The six columns are: (a) where the information was published – reference; (b) the patient group; (c) the patients problem or behaviour to be tackled; (d) the hypothesis or area investigated – for example colour impacting orientation; (e) the intervention proposed in an existing building; and (f) the findings of the intervention – that is whether there was improvement in patients' health and well-being. Although this might seem a very simplistic way of organising the information, the person responsible for including studies in the database or using the evidence should be aware of: (a) how outcomes were measured; (b) the relevance of the outcomes in improving patients' condition; and (c) the extent to which they were related to the design intervention proposed.

11.5.2 Framework 2: root cause and effects

The objective of developing a cause–effect evidence database is to inform designers and healthcare planners about the root causes of problems related to poor service and building design. In general, evidence is gathered from studies evaluating existing environments (e.g. the causes of noise in wards). Although those studies may not present solutions for the identified problems, they may lead designers to think of design solution that does not work.

Table 11.2   Patients group framework.

Patients group framework

(a) Source of evidence	(b) Illness and patients' condition	(c) Patients' problem to be solved	(d) Hypothesis/ area of investigation	(e) Intervention on building	(f) Effect on patient
Passini et al. (2000)	Dementia and Alzheimer	Way finding and disorientation	Visual stimulation through signage to support patients	Use of sign-posts and signage in the building	Personal autonomy and quality of life increase
Simpson et al. (2004)	Geriatric hip fracture	Fracture due to falls	Impact of different floor types in the risk of fractures	Changes on the type of flooring (wood and concrete base – carpeted and non-carpeted)	Risk of falling uniform in all floor types Better results for concrete carpeted
Wolfe (1975)	Children with psychiatric disorders	Increased agitation and anxiety levels	Room density and crowding	Resizing of the bedrooms	Improved privacy and behaviour

Table 11.3 presents data organised according to cause and effects. The columns are: (a) the source of evidence; (b) the built environment characteristic associated with a health problem (e.g. noise, light, occupancy); (c) the underlying causes of such problems, that is the root cause; and (d) the effects on health. The advantage of using this framework is that the identified causes can be common to different areas within a healthcare facility, for example noise in bedrooms, wards, corridors and waiting areas.

11.5.3 Framework 3: specific built environment characteristic framework – colour

In relation to the specific characteristics of the built environment, it is important to highlight that the impact on humans may be positive, negative or neutral. It is also important to identify that some characteristics may have several variants, such as colour and light for example. In the following framework variants in both the built environment characteristic and health outcomes are listed so that in the left-hand side negative impacts are listed whereas in the right-hand side positive ones are presented. Blank cells represent neutral effects on specific health outcomes and the numbers represent the source of evidence which is listed separately.

Table 11.4 presents the framework organised according to the rainbow colours. The advantage of using this framework is that an overall view about the impacts of colour can be formed. A limitation of this approach relates to the wide range of available colours and tones for which evidence might not be available.

Table 11.3 Root cause and effects framework.

Root cause and effects

(a) Source of evidence	(b) Built environment characteristic	(c) Underlying cause (root causes)	(d) Effects
Ersser *et al.* (1999); Bayo *et al.* (1995); Douglas and Douglas (2005)	Inappropriate levels of noise	Machines (e.g. haemodialysis machine); wheelchairs and stretcher locomotion; conversation between roommates or visitors; noise from outside; open plan; ward layout	Sleeplessness; Heart rate increase; Blood pressure increase; Anxiety; Stress increase;
Beral *et al.* (1982); Miller *et al.* (1995); Veitch and McColl (1993)	Inappropriate levels of natural and/or artificial light	Artificial fluorescent light; artificial cycled light; natural ultraviolet light;	Sleeplessness; seasonal affective disorder; melanoma
Altimier (2004); Baskaya *et al.* (2004)	Crowding	Room size; furniture layout; density	Stress increase; sleeplessness; anxiety

11.5.4 Framework 4: built environment and health outcomes[2] – overview

This framework was developed with the aim of increasing transparency in the dissemination of research results linking the built environment to patients' health outcomes. Rather than a descriptive approach, this framework uses a more visual approach which shows the variables investigated within a piece of research.

Basically, the identification of the main elements and variables within the selected papers were listed and classified in an excel spreadsheet. The list is not exhaustive, but the framework was developed to be incremental, so new variables can be added. Three main categories are considered within this framework: (a) patient profile (background), (b) the built environment, and (c) health outcomes. These categories are subdivided into other subgroups as describe in the following list.

- Patients' profile: This category relates to the information about the patient when the study was conducted and it is subdivided into four subgroups: (a) illness or patient group; (b) age group, that is infant, adult, older; (c) gender – male of

---

[2] This framework was developed as part of a research project within HaCIRIC (Health and Care Infrastructure Research and Innovation Centre) and used elements of systematic literature review. Here only an overview is presented and more information can be found at Codinhoto *et al.* (2008).

Table 11.4 The impacts of colour on health outcomes.

	Negative impacts							Health outcome	Positive impacts						
	V	I	B	G	Y	O	R		R	O	Y	G	B	I	V
Psychological								Depression							
					2		2	Anxiety				2	2		
								Stress							
								Insecurity							
								Fear							
								Panic							
								Confusion							
								Satisfaction							
								Well-being							
		4					1	Arousal				1, 4			
Physiological			3	3	3		3	Respiration	3		3	3	3		
								Coordination							
								Excretion							
								Circulation							
								Reproduction							
Physical			3	3	3		3	Heart rate	3		3	3	3		
								Pain							
								Hypothermia							
								Blood pressure							

1. Wilson (1966).
2. Jacobs and Suess (1975).
3. Jacobs and Hustmyer (1974).
4. Nourse and Welch (1971).

female; and (d) patient condition – this refers to whether or not the patient had an intervention (e.g. post-operation) or is about to have one (pre-operation);

- The built environment: In total, eight subcategories were utilised: (a) specialist, that is what type of healthcare facility the study is about – for example hospital, primary care, nursing home; (b) care units referring to specialised areas within the facility – for example cancer unit, intensive care unit, coronary care unity; (c) setting: this refers to the specific location of the investigation – for example bedroom, corridor, operation theatre; (d) component: this refers to the parts that define the environment – for example walls, ceiling, floor; (e) furniture and equipment; (f) systems – for example ventilation, lighting, sound; (g) functions: this refers to the use of the space – for example layout, safety, accessibility; and

(h) characteristics: this is the most detailed level of the framework and refers to the fabric and ambient environment parts and elements of the investigation – for example texture, colour, dimension, temperature;

- Health outcomes: This category describes health outcomes measured in the study and the results achieved (positive, negative and neutral). There are four subcategories considered: (a) psychological: referring to the problems of the mind – for example anxiety, arousal, depression and stress; (b) physiological: in relation to the internal systems of our body – for example circulation, respiration, reproduction; (c) physical: referring to specific measures of body integrity and health – for example heart hate, melanoma, fracture; and (d) Other health outcomes: this refers to indirect measures of health outcomes – for example healing time, length of stay and well-being.

In total 176 variables were considered in this framework. The benefit of using this framework relates to the fact that a map of the existing evidence is built and gaps in research can be identified. Additionally, in the case of lack of evidence in relation to a specific setting, studies conducted in similar environments can be used.

11.6 Conclusions

Evidence-based design is an approach to support decision making which emphasises the use of scientific evidence. The use of evidence is important for critical decisions where a set of vital information about the impact of design solutions upon users and maintenance may influence the way design evolves. Disconnected pieces of evidence should not be mistakenly used as EBD to justify bias within design solutions. Rather, evidence should support decisions and whenever possible, designers and healthcare planners should collect relevant information from completed projects in order to update the evidence-base. In other words, this means to check whether or not their decisions efficiently and effectively improved the quality and use of the space.

Currently there are limitations in maximising the utilisation of EBD. These are related to the lack of explicit cause and effect relationships, the fragmentation and sparseness of the available information and methodological limitations. However, the use of systematic reviews can mitigate those problems and bring strength to EBD. EBD is evolving fast with a rapidly growing body of evidence.

Moreover, the implications of EBD to the design process have not been deeply explored yet. However, issues related to changes in the configuration of the design team (e.g. by considering the participation of a researcher) and the provision of evidence to designers have started being explored. Furthermore, discussions about whether EBD aligns or contradicts new design and production theories and methods, such as lean design and production and the link between parametrical design and EBD are also emerging.

Finally, studies linking the built environment and effects on people involve a considerable number of variables that can be organised in different ways. In this chapter, several systematic steps for gathering evidence and different ways of compiling research findings are suggested. These frameworks address different issues from specific relationships to provide a broad overview of cause–effect relationships. These frameworks can be used by designers and healthcare professionals to support decision making through EBD. By adopting an evidence-based approach, clarity and transparency can be brought to the design decision-making process and EBD can support professionals making better-informed decisions.

11.7 Acknowledgements

This work was funded by the Engineering and Physical Sciences Research Council and conducted through the Salford Centre for Research and Innovation and the Health and Care Infrastructure Research and Innovation Centre.

References

Altimier, L. (2004) Healing environments: For patients and providers. *Newborn and Infant Nursing Reviews*, 4(2): 89–92.

Baskaya, A., Wilson, C., and Ozcan, Y.Z. (2004) Wayfinding in an unfamiliar environment: Different spatial setting of two polyclinics. *Environment and Behavior*, 36(6): 839–867.

Bayo, M.V., Garcia, M.A., and Garcia, A. (1995) Noise levels in an urban hospital and workers' subjective responses. *Archives of Environmental Health*, 50: 247–251.

Beral, V., Shaw, H., Evans, S., and Milton, G. (1982) Malignant melanoma and exposure to fluorescent lighting at work. *The Lancet*, 7: 290–293.

Boaz, A., Ashby, D., and Young, K. (2002) Systematic reviews: What have they got to offer evidence based policy and practice? Working paper 2, ESRC UK Centre for Evidence Based Policy and Practice, Queen Mary, University of London, http://www.evidencenetwork.org.uk/Documents/wp2.pdf

Codinhoto, R., Tzortzopoulos, P., Kagioglou, M., Aouad, G., and Cooper, R. (2008) The effects of the built environment on health outcomes. Research report, HaCIRIC (Health and Care Infrastructure Research and Innovation Centre). Available at http//www.haciric.org

Codinhoto, R., Tzortzopoulos, P., Kagioglou, M., Aouad, G., and Cooper, R. (2009) The impacts of the built environment on health outcomes. *Facilities*, 27(3/4): 138–152.

Cooper, R., Boyko, C., and Codinhoto, R. (2008) *The effect of the physical environment on mental wellbeing, State-of-Science Review: SR-DR2, Mental Capital and Wellbeing: Making the most of ourselves in the 21st century*. London: Government Office for Science.

Dennis, J.L. and Lomas, J. (2003) Convergent evolution: The academic and policy roots of collaborative research. *Journal of Health Services Research and Policy*, 8(2): 1–5.

Devlin, A.S. and Arneill, A.B. (2003) Health care environments and patient outcomes: A review of the literature. *Environment and Behavior*, 35(5): 665–694.

Douglas, C.H. and Douglas, M.R. (2005) Patient-centred improvements in health-care built environments: Perspectives and design indicators. *Health Expectations*, 8: 264–276.

Ersser, S., Wiles, A., Taylor, H., Wade, S., Walsh, R., and Bentley, T. (1999) The sleep of older people in hospital and nursing homes. *Journal of Clinical Nursing*, 8: 360–368.

Jacobs, K.W. and Hustmyer Jr., F.E. (1974) Effects of four psychological primary colors on GSR (galvanic skin response) heart rate and respiration rate. *Perceptual and Motor Skills*, 38: 763–766.

Jacobs, K.W. and Suess, J.F. (1975) Effects of four psychological primary colors on anxiety state. *Perceptual and Motor Skills*, 41: 207–210.

Johnson, P. and Duberley, J. (2006) *Understanding management research*. London: SAGE Publications Ltd.

Landry, R., Amari, N., and Lamari, M. (2001) Utilization of social science research knowledge in Canada. *Research Policy*, 30: 333–349.

Lord Hunt of Kings Heath. (2002) Improving the patient experience: The art of good health – A practical handbook. NHS Estates London: TSO, p. 4.

Malkin, J. (2008) *A visual reference for evidence-based design*. California, USA The Center for Health Design.

Miller, C.L., White, R., Whitman, T.L., O'Callaghan, M.F., and Maxwell, S.E. (1995) The effects of cycled versus non-cycled lighting on growth and development in pre-term infants. *Infant Behavior and Development*, 18: 87–95.

Mulrow, C.D. (1994) Systematic reviews: Rationale for systematic reviews. *BMJ*, 309(6954): 597–599.

Mulrow, C.D. and Cook, D.J. (1997) Formulating questions and locating primary studies for inclusion in systematic reviews. *Annals of Internal Medicine, Academia and Clinic, Systematic Review Series*, 127(5): 380–387.

NHS Estates. (2005) Safer Environment Database – EFM-Evidence. Lawson, B. and Phiri, M.

Nourse, J.C. and Welch, R.B. (1971) Emotional attributes of color: A comparison of violet and green. *Perceptual and Motor Skills*, 32: 403–406.

Passini, R., Pigot, H., Rainville, C., and Tetreault, M.H. (2000) Wayfinding in a nursing home for advanced dementia of the Alzheimer's type. *Environment and Behavior*, 32(5): 684–710.

Russell, B. (1948) *Human knowledge: Its scope and limits*. New York: Simon and Schuster, 524p.

Sackett, D.L. (1997) Evidence-based medicine. *Seminars in Perinatology*, 21(1): 3–5.

Simpson, A.H.R.W., Lamb, S., Roberts, P.J., Gardner, T.N., and Grimley Evans, J. (2004) Does the type of flooring affect the risk of hip fracture. *Age and Ageing*, 33: 242–246.

Smith, T. (2003) *An evaluation of sorts: Learning from common knowledge*. Durham: CAHHM, University of Durham, p. 21.

Staricoff, R.L. (2004) Arts in health: A review of the medical literature. Research report 36, Arts Council England, London.

Sundstrom, E., Bell, P.A., Busby, P.L., and Asmus, C. (1996) Environmental psychology 1989–1994. *Annual Review Psychology*, 47: 485–512.

Tranfield, D., Denyer, D., and Smart, P (2003) Towards a methodology for developing evidence-informed management knowledge by means of systematic review. *British Journal of Management*, 14: 207–222.

Ulrich, R., Quan, X., Zimring, C., Joseph, A., and Choudhary, R. (2004) *The role of the physical environment in the hospital of the 21st century: A once-in-a-lifetime opportunity*. Concord, CA: Center for Health Design for Designing the 21st Century Hospital Project, p. 69.

Ulrich, R., Zimring, C., Zhu, X., DuBose, J., Seo, H., Choi, Y., Quan, X., and Joseph, A. (2008) A review of the research literature on evidence-based healthcare design. *Health Environments Research & Design Journal*, 1(3): 61–125.

Veitch, J.A. and McColl, S.L. (1993) *Full spectrum fluorescent lighting effects on people: A critical review (no. 659)*. Ottawa, Canada: Institute for Research in Construction.

Wilson, G.D. (1966) Arousal properties of red versus green. *Perceptual and Motor Skills*, 23: 942–949.

Wolfe, M. (1975) Room size, group size, and density: Behavior patterns in a children's psychiatric facility. *Environment and Behavior*, 7: 199–224.

Benefits Realisation

12

Planning and Evaluating Healthcare Infrastructures and Services

Stelios Sapountzis, Kathryn Yates, Jose Barreiro Lima and Mike Kagioglou

12.1 Introduction

Traditionally, investment programmes (and projects[1]) determine their level of success mainly against cost, quality and time of delivery, and not on a full characterisation of benefits or impacts delivered. Targeting clarification of impacts and benefits, known as benefits realisation, is emerging as a method to assist organisations to manage whole life cycle of programmes (Glynne, 2007), from development, construction and facilities management, to operations management and back-office services delivery.

Following the emergent importance of benefits realisation applied to healthcare infrastructures and services, the Health and Care Infrastructure Research and Innovation Centre (HaCIRIC) based in the UK is undertaking a research initiative targeting the development of a benefits realisation model (BeReal model). BeReal

---

[1] In this chapter the terms programme and project are used alternatively, although it is recognised that project represents the overall scope of work being performed to complete a specific job, typically under a temporary endeavour and to create a unique product or service (Phillips, 2004); whilst programme is understood as a portfolio of projects selected, planned and managed in a coordinated way.

focuses on how benefits should be elicited at the initial/strategic stages of a programme, how benefits should be managed and traced along the life cycle and deployed within the programme's business case. Subsequently, BeReal aspires to be an appropriate method to drive and control the programme plan; providing tools and techniques for defining specific benefits. It also allows the measurement and evaluation of the extent to which those benefits are delivered (or not).

This chapter's objectives are:

- to present how benefits are widely understood and used within the context of major investment programmes;
- to justify the need for benefits realisation (and management) as a planning, evaluating, optioning and control method within the overall context of healthcare infrastructures (e.g. buildings, access) and services (e.g. hard and soft services to healthcare facilities);
- to introduce a high-level view of the BeReal model;
- to provide multi case-study contextual perspectives, highlighting central aspects related to the high-level version of BeReal model development and its future application.

12.2 Benefits realisation

Benefits management is defined as the process for the optimisation of benefits from a programme's change perspective (Reiss *et al.*, 2006). Nevertheless, the concept of benefits realisation is not new (Simon, 2003, in Nogeste, 2006). The existent literature on benefits realisation consists of practical guides and frameworks around information technology/information systems (IT/IS) project investment justification within the private sector. More recently, benefits realisation (and management) has been raised as the 'new' practice for private and public sector programmes in a diversity of sectors, including healthcare infrastructure, housing developments and education. Benefits realisation is the evolution from the general investment programme appraisal approach (usually based on cost, quality and time), to an active planning approach supported by benefits planned, delivered and realised (or not) by stakeholders (Glynne, 2007).

Healthcare programmes are complex systems (Carruthers *et al.*, 2006; Sweeney and Griffiths, 2002) developed within multi-stakeholder environments. Such programmes typically have a long lifespan, with many phases including policy setting, planning, development, construction, commissioning of healthcare service operations, facilities management (e.g. maintenance), refurbishments and demolition. The complexity induced by stakeholders diversity and the long lifespan of programmes may induce failure in delivering aims set out in its initial stages, that is stakeholders are not realising what has been planned. Disparity or lack of correspondence between the *ex-ante* and *ex-post* stages

(Farbey *et al.*, 1999; Silva, 1995) can be related to poor (or lack of) benefits management. Benefits realisation supports that such disparity between life cycle stages is managed through a process that involves eliciting, managing, monitoring/controlling, realising and measuring benefits.

Benefits realisation is particularly important within healthcare environments as the process 'helps to ensure a clear sign posting of who is responsible for the delivery of those benefits' (NHS, 2007). Furthermore, the continuous changes in the structure and governance of the National Health Service (NHS) make it difficult to evaluate services, benefits realisation would enable these important evaluations to occur under a unique and progressively optimised approach.

The BeReal model presented in this chapter aims to introduce an approach that will assist to identify, manage and monitor benefits throughout a programme's life cycle by providing facilitation for evidence-base decision-making, continuous improvement and organisational learning.

12.2.1 Benefit taxonomies

McCartney (2000, p. 35), in the modernising government in action report, states that: 'Projects and programmes can only be regarded as successful if the intended benefits are realised'. Furthermore, a common characteristic of many unsuccessful programmes is the vagueness with which the expected benefits are defined (Reiss *et al.*, 2006). Prior to defining the programme's objectives, a clear understanding is needed about which strategic benefits should be considered and how benefits should be understood as drivers in achieving objectives. Considering the need of a clear understanding about *benefits*, a brief summary of the literature in terms of benefits definitions and benefits organisation (i.e. classification and characterisation) is provided in Section 12.4.5.

Benefit – the word benefit is widely used and very broadly defined as 'a measurable improvement'. A benefit is an outcome whose nature and value are considered advantageous by an organisation (Ward *et al.*, 1995). Benefits should be *owned* by stakeholders who want to obtain value from an investment (Glynne, 2007; Ward and Daniel, 2006).

Dis-benefits – Ward *et al.* (1996) highlight that potential dis-benefits of an investment should always be considered, defining those as adverse impacts on a business and/or organisation. An organisation or investors need to agree that negative outcomes or dis-benefits are the potential 'price worth paying to obtain positive benefits' (Ward *et al.*, 2004). Some *outcomes* maybe favourable for the organisation as a whole but unfavourable for parts of it, so any such dis-benefits must be identified and tracked so that their impact can be minimised (CCTA, 1999).

Segmentation of (dis-)benefits is approached by many authors, considering different aspects or criteria; segmentation helps to increase the understanding of the nature of benefits, and assists analysis and communication (Bradley, 2006). Table 12.1 contains examples of segmentation types, being the most common distinction made between *tangible* and *intangible* benefits.

Table 12.1 Benefits segmentation – selected examples.

Aspects	Comments
Tangible, intangible (Bradley, 2006; Ward and Daniel, 2006)	Tangible are those judged *more* objectively, considering quantitative measures which are often but not always financial (i.e. in monetary values). Intangible are those judged *more* subjectively, most often measured employing qualitative measures (note: *intangible* benefits are generally difficult to convert to monetary values).
Organisational impact, business impact (Farbey et al., 1999)	These benefits can come in five different business streams: (1) strategic, (2) management, (3) operational, (4) functional and (5) support.
Built environment life cycle view (Yates et al., 2009)	The built environment life cycle states that views might be segmented as follows: (1) development, (2) construction and (3) facilities management.
Stakeholder, actor orientated (Mantzana and Themisto-cleous, 2004)	Classification according to the stakeholder (groups) who will feel or experience impacts. In an investment, project or programme the actors/stakeholders might be classified in four main categories, these are: (1) providers, (2) acceptors, (3) supporters and (4) controllers. All these four main categories might also be classified in two different dimensions: (1) human and (2) organisational.
Planned, unplanned/emergent (Ashurst and Doherty, 2003)	Unplanned benefits are often a complementary consequence of (other) planned benefit(s) and/or of an implemented change(s); due to the emergent nature of benefits, these are not usually documented in the business case, not being initially related with planned changes and/or investments detailed.
Organisational view (Yates et al., 2009)	Organisational views are: (1) operations management and (2) back-office, which encloses as main functions (2.1) information systems, (2.2) human resources, (2.3) legal, (2.4) purchasing and procurement, (2.5) finance and accounting and (2.6) other office services.

12.3 Research methodology

Central to the research methodology was the choice of epistemology for the study; it is from here that choices relating to research techniques and approaches could be made. As the research is focused on the real world and what is happening within it, the research philosophical choice for this was based on interpretivism, focusing upon observation and description of phenomena (Silverman, 2006). Interpretive research uses the interpretations of social actors to understand that phenomenon (Sonali and Corley, 2006). Interpretive data analysis should provide the researcher with rational and credible understanding of a phenomenon.

Along the next paragraphs, major methodological research choices are briefly summarised.

12.3.1 Research strategy

The research strategy chosen was action research and case study. Case studies can be used in the preliminary stages of an investigation to generate hypotheses, as well as testing out hypotheses and phenomena in the real world (Flyvbjerg, 2007; Yin, 2003). The majority of healthcare programmes have a life span of 20–30 years, which presents a constraint in choosing a single project to act as a case study in developing, implementing and validating the BeReal model. Therefore, multiple case studies were conducted, using projects at different phases of their life cycle. The use of multiple case studies makes cross-validation possible and increases the strength of the study by enabling the findings from each case to be evaluated against the others (Herriot and Firestone, 1983).

As summarised in Table 12.2, coverage of the identified generic phases of a healthcare programme is comprehensive in terms of the built environment life cycle. *Policy setting*, *programme development* and *business case approval* (BCA) are considered within the *development* phase. After *construction*, *post-project/ occupancy evaluation* and *operational* programme phases are considered within the *facilities* dimension and the *operations and back-office* view (organisational view). Evaluations have occurred over the same time period but with different cases, which means that the study is developed under a cross-sectional (Easterby-Smith *et al.*, 2002) approach.

Action research allows researchers to work actively with industry so that the needs of both groups are addressed through the research; and if needed cause a change within a social system (Lewin, 1946). In this research, action research follows the 'regulative cycle' of: research question, diagnosis, plan, intervention

Table 12.2 NHS programme phases, built environment life cycle and organisational views.

| Generic healthcare programme phases | Built environment life cycle view | | | Organisational view |
	Development	Construction	Facilities	Operations and back-office
Policy setting (PS)	✓			
Programme development (PD)	✓			
Business case approval (BCA)	✓			
Construction (CON)		✓		
Post-project/occupancy evaluation (POE)			✓	✓
Operational (OP)			✓	✓

Legend: ✓, Main focus.

and evaluation (van Strien, 1975) through interaction between researchers and industry. Through this interaction and continued discourse between the two groups a user community, in this case a user community of BeReal is developed, this community provides direction and advice to the researchers, whilst the researchers provide an intervention (the BeReal tool), for the problem diagnosed by the group and in literature (business cases are used as tick boxes to get money and rarely revisited).

12.3.2 Research techniques

Both quantitative (questionnaires) and qualitative (interviews, stakeholder workshops) techniques were used. Such triangulation strengthens the validity and accuracy of the research (Yin, 2003), and the attributes of each technique can complement the other (Punch, 2005).

Questionnaires were used to provide a relatively quick and low-cost way to gain a reliable insight into 'characteristics, attitudes, and beliefs' (Marshall and Rossman, 1999; May, 1993). For the Manchester, Salford and Trafford (MaST) Local Improvement Finance Trust (LIFT) case study, questionnaires were used to gain an insight and assess the different groups' views and perceptions of the services, facilities and overall effects of the LIFT scheme. Questionnaires were chosen as they can be used on a small population sample, and results used to make inferences about that population (Holton and Burnett, 1997). Also, patterns and causal relationships can be identified from the data gathered.

Semi-structured interviews have been used to collect qualitative data, as they provide guidance through some predetermined questions but also allow flexibility, as the questions can alter depending on the answers given and how the interviewer/ researcher interprets the conversation (Robson, 2003). These also allowed the researchers to delve deeper into some issues arising from questionnaires and patient forums.

Finally, most of the data collected in the project was through workshops with stakeholders. These were similar to focus groups, being used to gain the views and perceptions of stakeholders. Indeed, patient and public involvement forums are already used in the NHS to 'Monitor and review NHS delivery, seek the views of the public about those services, make recommendations to the NHS accordingly and other issues relevant to organisation' (DoH website, 2008).

From the different techniques emerged quantitative and qualitative data to be analysed, both primary (obtained from the questionnaires, stakeholder workshops and interviews) and secondary data (obtained from the data collection of Primary Care Trust (PCT) records). Analytical software will be used to analyse the data collected; Non-numerical, Unstructured Data Indexing, Searching and Theorising (NUD*IST) software for qualitative data and Statistical Product and Service Solutions (SPSS) for quantitative data.

NUDIST software is beneficial when looking into large descriptive texts (Yin, 2003) that may have resulted from the semi-structured interviews or workshops as it is based on a code-and-retrieve facility. This facility codes large amounts of data, which allows qualitative matrices, relationships and patterns to be identified (Richards and Richards, 1994). From the findings NUD*IST can be used to develop and influence new ideas and hypothesis.

SPSS can be used to undertake statistical analysis from which analytical reporting and graphics can be produced, and regression analysis, factor analysis, correlations, analysis of variance and reliability testing can be carried out.

12.4 BeReal model overview

The BeReal model presents four non-sequential phases, illustrated in Figure 12.1 and described in the following paragraphs. It classifies benefits into a three-level hierarchical organisation. The four main phases of the BeReal were defined through an extensive literature review, informed by benefits elicitation workshops with the participation of diverse stakeholders' groups.

12.4.1 Phase 1: benefits management strategy and benefits realisation case

This phase is concerned with identifying and documenting desired *strategic benefits* and *sub benefits* and developing a benefits management strategy to share and communicate these to relevant stakeholders.

Bennington and Baccarini (2004) suggest that benefits elicitation should emerge from a combined approach of interviews and workshops involving key stakeholders. Remenyi and Smith (1998) argue that a key aspect of the benefits identification/ elicitation process is that stakeholders better understand the investment, what is affordable and possible. The involvement of key stakeholders identifying and

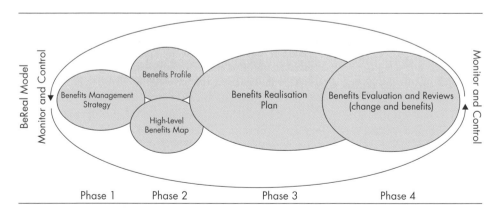

Figure 12.1 The BeReal model.

agreeing on desired benefits is essential as it enhances awareness and maximises the likelihood of commitment to deliver and realise those benefits across a range of levels in the business and in the organisation (Glynne, 2007; Kagioglou et al., 2000). Stakeholders will need to understand how benefits are to be identified, modelled and subsequently delivered (Reiss et al., 2006). Nevertheless, although it would be risky to assume that all stakeholders will understand the implications of benefits identification and planning, involving the stakeholders along the process may help to foster a team environment and encourage appropriate communication and enable better decision-making.

12.4.2 Phase 2: benefits profile and high-level benefits map

Profile and mapping are two major enabling tools. These are powerful as they hold core information of each individual benefit, and might support illustration of dependencies/interrelationships and overlaps between (dis-)benefits. Investing time at initial stages to establish robust and reliable *benefit profiles* and *maps* enables management of the benefits realisation life cycle.

Benefit profile – Many authors (Bartlett, 2006; Bennington and Baccarini, 2004; Bradley, 2006; CCTA, 1999; Farbey et al., 1999; Reiss et al., 2006; Thorp, 1998; Ward and Daniel, 2006) recommend drawing up formalised benefit profiles so that they can be managed with a similar rigour as costs of projects.[2] Nevertheless, along with other aspects *profile*, *dependencies* are not understood as static; for example, as the programme progresses, the benefits and the benefit profile information will suffer revisions (CCTA, 2000). Benefit profiles should include a number of aspects, for example:

- description for each benefit or dis-benefit;
- how it will be measured (e.g. formula, source of data);
- qualitative or quantitative benefit (note: financial valuation is possible/ difficult);
- how it will change current business processes and functions;
- how it interacts with other benefits (i.e. dependencies/interrelationships);
- the extent to which it depends on the success of other ongoing/future projects.

Mapping – The benefits dependency network (or benefits mapping) was first introduced by Ward and Elvin (1999) and aims to illustrate potential 'many to many' relationships, as for example between enabling changes, business changes, business benefits and investments' objectives. A benefits dependency network,[3] targets

---

[2] The premise is that under a *cost–benefit* approach, *costs* are more easily quantified than *benefits*, although *benefits* are usually linked with *costs*.
[3] A *benefits map* example is given in Figure 12.7.

to map all the *cause and effect* relationships, and to include stakeholders, changes and criteria for success (Bradley, 2006).

Elicitation of *strategic*, *sub benefits* and *end benefits*, development of benefits profiles and design of the benefits map will form the basis of necessary ongoing update of the benefits realisation plan.

12.4.3 Phase 3: Benefits realisation plan

Focus is on the development and execution of a benefits realisation plan informed by the previous phases content/data, consisting of measuring and tracking the benefits previously identified and incorporating emerging benefits.

OGC (2007a) underlines the importance of doing a management plan that describes how the organisation wishes to manage and achieve benefits, from any investment in business change. The benefits realisation plan developed during phase 3 (that includes key assumptions, risk assessment and sensitivity analysis related to those *outcomes* that contribute most to expected benefits) should be seen as a major tool contributing to the decision-making process (OGC, 2007b). Without this plan it is difficult to predict how an organisation might effectively realise benefits (Ward *et al.*, 1996).

12.4.4 Phase 4: benefits evaluation and reviews (change and benefits)

This phase includes the evaluation and controlling of benefits, as these have been elicited through the previous phases and might be further identified, detailed and monitored during the post-project evaluation phase. Nevertheless, this phase might be initiated previously; since dis-benefits (and benefits) might emerge during the implementation management (e.g. construction works). For example, rehabilitation and refurbishments works, expansion of existing healthcare infrastructures and/or temporary changes to healthcare services due to construction onsite ongoing works, typically involve dis-benefits.

Benefits review is 'the process by which the success of the project is addressed, opportunities for the realisation of further benefits are identified, lessons are learned and opportunities for improvement in future projects are identified' (Ashurst and Doherty, 2003). Monitoring and controlling should be developed continuously, not only along the overall built environment life cycle, but also covering operations management services. Indeed, the benefits stated in a programme's business case are the result of an often protracted period of benefits planning. Furthermore, by the time the business case has been signed off, the benefits as originally conceived may have been significantly diluted (Bartlett, 2006).

A successful benefits management life cycle needs to include periodic reviews to reconfirm the alignment of the programme with the organisations strategic priorities. These reviews should also assess whether or not the anticipated suite of benefits is sufficient to accomplish the organisations' strategic goals by also contacting benefits reviews at project stage gates[4] (Cooke-Davies, 2002; Nogeste, 2006).

According to Ward *et al.* (1996), specific responsibility for realising the benefits is allocated within the business and/or within the programme for each benefit. When identifying roles the task should include the stakeholders affecting the delivery of each benefit, and the changes and tasks needed to ensure delivery (Sakar and Widestadh, 2005). Responsibility and leadership for benefits management falls into three main areas and professional categories (CCTA, 1999):

- Identifying and defining: Business Change Manager;
- Planning and monitoring: Programme Support Officer/Programme Manager;
- Realisation: Line Manager.

Cooke-Davies (2002) in Nogeste and Walker (2005) further highlights the importance of governance and proposes that the following three principles underpin any effective benefits management system:

- Create project governance structures that involve both the project structure and the functional lines of the organisation;
- Drive all governance decisions about the project through the business case;
- Redefine project management methods and frameworks so as to make benefits management an integral part.

12.4.5 BeReal usability and controlling structure

BeReal organises benefits in three main categories, represented in Figure 12.2.

Strategic benefits are related with the purpose of characterising the programme, providing an overall direction of success throughout the life cycle; *sub benefits* characterise specific targets linked to strategic benefits and should drive design and preliminary evaluation of alternatives of investment (e.g. selection of building options); *end benefits* are measures that characterise in detail (e.g. hard, soft, tangible, intangible, quantitative, qualitative) the targeting and achieving of benefits – see Figure 12.3.

The BeReal controlling structure is designed to cover the overall *investment life cycle* enclosing the *built environment* and *organisational views*, supporting (1) different measurement levels depending on the *characterisation* of benefits, and (2) providing consistent deviation control between '*actual (as-is)*', '*planned (to-be)*'

---

[4]*Stage gate review* is a generic term meaning a point at the end of a phase where a decision is made whether a project should continue (OGC, 2005), being fully confirmed without changes, partially changed or globally terminated.

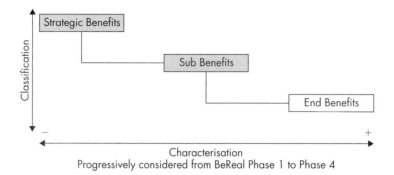

Figure 12.2   BeReal model – benefits organisation.

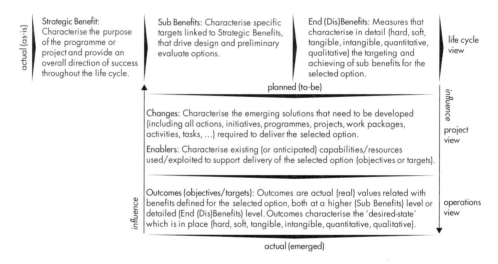

Figure 12.3   BeReal model – controlling structure.

and '*actual (emergent)*' investment life cycle moments. Table 12.3 presents an overview of the BeReal controlling structure.

Actual versus estimated[5] – The '*actual (as-is)*' data is related with what is measured and is being *actually* realised (or not) on the contextual *status quo*, prior to the investment. The '*planed (to-be)*' emergent data is related with (what is being planned and) what is being *estimated* and should be realised in the future. '*Actual (emergent)*' data relates with what is actually being measured and is being *actually*

---

[5] Alternatively and as proposed by Farbey *et al.* (1999) benefits might be considered *ex-ante* or *ex-post*, depending on the timing in which benefits are measured (and not so much related with when the benefits are elicited); generally benefits are planned considering *ex-ante* data (*real* and *targeted*) and evaluated against *ex-post* (real) data.

Table 12.3 BeReal phases and benefits organisation.

BeReal phases		Benefits organisation		
Descriptions	**Phases**	**Strategic**	**Sub**	**End**
Benefits management strategy	I	✓		
Benefits profile	II	✓	✓	
High-level benefits map	II	✓	✓	
Benefits realisation plan	III	✓	✓	✓
Benefits evaluation and reviews (change and benefits)	IV	✓	✓	✓

Legend: ✓, Main focus.

delivered/realised to/by the stakeholders. In order to assure a controlling approach along the investment life cycle phases, consistency between phases and the different benefits organisation levels ('as-is', 'to-be' and 'emergent') is understood as critical.

12.4.6 Investment appraisal approaches: general, healthcare specific and BeReal model

As discussed initially, generally investment programmes determine their level of success mainly against cost, quality and time of delivery, and not based on a full characterisation of benefits or impacts delivered. One of the leading general decision-making approaches for appraising major capital investment programmes used across different sectors and recommend by the UK government is the Gateway© Review by the Office of Government Commerce (OGC). The Gateway Review highlights the need for a robust benefits realisation methodology when managing programmes.

Along with the review and for the purposes of this chapter the BeReal model is further aligned with two traditional investment approaches currently used within UK's healthcare sector (see Figure 12.4):

- Capital Investment Manual (CIM) used by the Department of Health (DoH);
- Traditional private finance initiative (PFI) investment.

At this stage of the research, the BeReal model does not intend to replace a healthcare organisation's investment process but to run alongside and help those involved to understand how the focus on benefits can further assist the management of the programme.

The case studies presented in the Section 12.5 have been selected aiming to validate the BeReal model and to also highlight the relevance of the BeReal phases with the corresponding traditional investment process stages. This dual approach, that mixes an existing (and stakeholders' known) model with a new and emerging

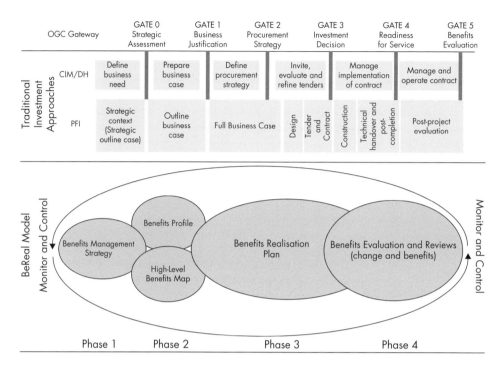

Figure 12.4 Investment appraisal approaches: general, healthcare specific and BeReal model.

model, can further enable BeReal's use as a tool that brings individuals and teams together whilst breaking down organisational barriers by shifting people's thinking towards benefits and outcomes.

12.5 Case studies

This section summarises major findings of two case studies, used to inform and validate the BeReal model. Major factors supporting selection of these cases studies included here (amongst other ongoing case studies available) are:

- *Overall view and case studies interrelation.* The fact that MaST LIFT focuses on the post-occupancy evaluation (POE) phase and Brighton and Sussex Tertiary, Trauma and Teaching (3Ts) at the BCA phase, was initially regarded and was afterwards confirmed as an opportunity to develop and further validate the BeReal model both on the initial phases more related with the benefits elicitation (i.e. *Phase 2: Benefits Profile and High-level Benefits Map*) and with the definition of a benefits realisation plan (i.e. BeReal model phase 3), until the operations and facilities management phase which is more related with delivering monitoring/controlling of benefits (i.e. *Phase 4: Benefits Evaluation and Reviews (Change and Benefits)*).

- In other words, the mix of these two case studies highly contributed to incorporate a short-term perspective related with the Brighton and Sussex University Hospitals (BSUH) 3Ts new development, with a more medium/long-term perspective related with the MaST LIFT.
- The above-mentioned and others aspects, as main methodological approaches used within each of the two case studies, are summarised in Table 12.4, schematically illustrated in Figure 12.5 and further discussed/detailed along the following sections.

Table 12.4   Selected case studies – characterisation summary.

Summary of aspects	Selected case studies	
	BSUH 3Ts	**MaST LIFT**
Generic healthcare programme phases[a]	BCA	POE
Built environment life cycle view[a]	Development	Facilities
BeReal regarded as a complementary model (see Section 12.4.6)	Yes	Yes
BeReal model phase[b]	Phase 2 and 3	Phase 4
Categories of stakeholders involved	>5[c]	>5[d]
Meeting attendees (overall no.)	48	16
Duration	4 months	12 months
Location	Brighton	Salford
Meeting facilitator	HaCIRIC	NHS
Stakeholders' meetings month/year	09/2008, 12/2008	12/2006, 01/2007
Elicitation source (research)	Meetings	Documentation
Overall no. of elicited benefits[e]	682	41
No. of elicited strategic benefits	8	5
No. of elicited sub benefits	37	36

References: Adapted from Yates et al. (2009); for further details on MaST LIFT Case Study consider Sapountzis et al. (2008).
[a]Please see Table 12.2.
[b]Please see Figure 12.1.
[c]Brighton and Sussex 3Ts Case Study examples of categories of stakeholders involved: (1) imaging, (2) cancer, (3) HIV/infectious diseases, (4) medicine/elderly care, (5) trauma/critical care/neurosciences, (6) programme board and (7) patient representatives. The participation of many other categories of stakeholders is scheduled/considered for subsequent phases of the programme.
[d]MaST LIFT Case Study major examples of categories of stakeholders involved during the elicitation meetings: (1) MaST LIFT Co Chief executive (2) Manchester PCT Finance Director, (3) Health Centre Managers, (4) Department of Health Gateway reviewer, (5) Manchester City Council, (6) Health Joint Unit Program Manager, (7) Primary Plus Facilities Manager and (8) Community Health Action Partnership Director. Many other categories of stakeholders were involved in other programme phases/stages, e.g. Community Health Partnerships. [e]For further information on benefits organisation, please consider Figure 12.2.

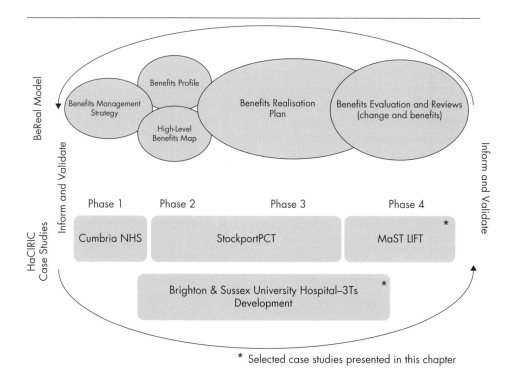

Figure 12.5 Selected cases studies alignment to inform and validate the BeReal model.

- *Study of a diversity of (focused) aspects.* Although the two case studies apparently have similar high-level patterns (see Figure 12.5, where major BeReal phases considered by each case study are highlighted), the idea of developing action research focusing on different/specific detailed aspects in each of the case studies was an excellent emerging opportunity to further develop and validate diverse components of the BeReal model.

Following the presentation of the above-mentioned factors supporting selection of these cases studies, the highlighted aspects identified are further discussed and detailed along the following sections:

Aspects discussed in both case studies

- Development of a clear benefits organisation vision, under a characterisation and classification approach (introduced in Figure 12.2).
- Active systematic organisation of the identified/elicited benefits, under a three-level organisation (summarised in Table 12.3).
- Benefits elicitation meetings with a diversity of stakeholders as a critical aspect.

Aspects discussed in one of the case studies

- Characterisation and test of a benefits weighting approach (summarised in Table 12.5).
- Characterisation and test of a benefits optioning approach (summarised in Section 12.5.1).
- Conceptual design of a monitoring/controlling structure covering the overall built environment life cycle and organisational views (firstly introduced in Table 12.1 and Section 12.3).

Table 12.5 BSUH 3Ts Case Study – identification, organisation and weighting of strategic and sub benefits.

Benefits[a]	Benefits classification[a]		Weighting % [b]
	Strategic benefits	**Sub benefits**	
1. Strategic fit (and contextual)	✓		**13.4**
1.01 Stakeholders alignment		✓	2.6
1.02 Synergy of services		✓	2.6
1.03 Context development		✓	2.6
1.04 Co-location/distribution		✓	3.7
1.05 Image, reputation, objectives		✓	1.9
2. Clinical outcomes	✓		**17.9**
2.01 Co-location		✓	3.0
2.02 Reduce referrals		✓	2.3
2.03 Improved quality of care		✓	6.3
2.04 Improve care outcomes		✓	6.3
3. Appropriate facilities (and facilities management)	✓		25.0
3.01 Fit-for-purpose building and infrastructure		✓	3.0
3.02 Facilities flexibility and future proofing		✓	3.0
3.03 Physical distribution of service locations (layout)		✓	4.0
3.04 Improved support services		✓	2.0
3.05 Increased patient/user safety		✓	5.0
3.06 Greater privacy (by better design)		✓	3.0

(Continued)

Table 12.5 (Continued).

Benefits[a]	Benefits classification[a]		Weighting %[b]
	Strategic benefits	Sub benefits	
3.07 Removal of backlog maintance		✓	1.0
3.08 Better working environment		✓	4.0
4. Access to services	✓		**10.7**
4.01 Service diversity/capacity fit		✓	5.4
4.02 Increased physical access		✓	3.5
4.03 Increased availability of services		✓	1.8
5. Training, teaching and research	✓		**8.9**
5.01 Improved research capability		✓	3.6
5.02 Improved teaching		✓	3.6
5.03 Knowledge transfer		✓	1.7
6. Use of resources	✓		**10.7**
6.01 Better equipment/recourses (technology)		✓	3.5
6.02 Better personnel		✓	1.8
6.03 Improved efficiency		✓	5.4
7. Operations management			**12.5**
7.01 Improved service coordination			3.5
7.02 Preventive health services			4.7
7.03 Improved user experience			4.3
8. Development and implementation	✓		**0.9**
8.01 Investment/change management effort		✓	0.9
8.02 Construction negative impact		✓	n/w
8.03 Planning ability		✓	n/w
8.04 Sustainability		✓	
8.05 Development feasibility		✓	n/w
8.06 Reduce service interruption		✓	n/w
8.07 Faster delivery (up to operation)		✓	n/w

References: [a]Sapountzis et al. (2009) and Yates et al. (2009). [b]Codinhoto and Passman (2009).

12.5.1 Brighton and Sussex University Hospitals Tertiary, Trauma and Teaching, case study

Tertiary, Trauma and Teaching is a hospital development programme by BSUH whose vision it is to provide clinical services, buildings and infrastructure that will be used by the local populations of Mid Sussex, Brighton and Hove for the next 30–40 years (Brighton and Sussex University Hospitals, 2009). BSUH 3Ts was at the BCA stage in May 2009, and has been through a number of other stages since 2007, as represented in Figure 12.5.

The principal aim of this case study was to elicit, classify and characterise benefits for the 3Ts hospital development as well as validating the BeReal methods of doing so. Furthermore, the case study aimed to develop and test methods for benefits weighting that were subsequently used for selecting between design options for the hospital development. These activities are in alignment with BeReal model phase 2 as illustrated in Figure 12.5. These methods and outputs of the case study are further discussed in detail in the following paragraphs.

Strategic benefits elicitation – In a workshop prior to the approval of the *Strategic Outline Case*[6] (SOC) six *strategic benefits* had been elicited: (1) *strategic fit*, (2) *clinical outcomes*, (3) *modern healthcare facilities*, (4) *improved access*, (5) *teaching, training and research* and (6) *effective use of resources*. Nevertheless, this group enlarged to eight strategic benefits, based on data that emerged from the (elicitation) Benefits Criteria Workshop with stakeholders (Figure 12.6).

Benefits elicitation – The initial activity that elicited benefits was based on a workshop with major patient stakeholder group members. The patient forum was facilitated by the BSUH 3Ts Project Director and two service improvement facilitators; 18 patient representatives were invited. The forum gave patients representatives an opportunity to discuss a number of issues including the design of the building, out-patient communications, transport, visiting, in-patient stay and leaving the hospital. From these discussions a total of 280 benefits were identified.

Benefits classification – This workshop was facilitated by the 3Ts Project Director and was attended by the relevant BSUH 3T stakeholders,[7] and occurred between the Benefits Identification workshop and the 1st Optioning workshop. This initiative has been aligned with BeReal phase 2, since it represents an opportunity to (BSUH 3Ts documentation, 2008) develop the following activities (not exhaustive):

---

[6] *SOC* is the first stage in the NHS approval process (Health Investment Plan) for projects that involve major capital investment (NHS, 2009); the document should be prepared by all the local NHS bodies, containing an initial assessment for a capital investment bid.

[7] Stakeholders: Director of Intensive Care, Consultant Histopathologist, Consultant Surgeon, Clinical Director, Emergency Care, PLC ENT/MFU/Urology/Breast, 3Ts Deputy Clinical Lead, Consultant Anaesthetist, Chief Nurse, Sexual Health & HIV consultant, Matron, a Medical Director and Project Manager (i.e. Cyril Sweett).

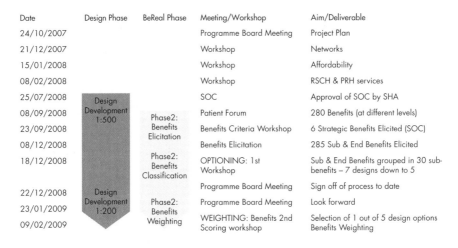

Date	Design Phase	BeReal Phase	Meeting/Workshop	Aim/Deliverable
24/10/2007			Programme Board Meeting	Project Plan
21/12/2007			Workshop	Networks
15/01/2008			Workshop	Affordability
08/02/2008			Workshop	RSCH & PRH services
25/07/2008	Design Development 1:500		SOC	Approval of SOC by SHA
08/09/2008		Phase2: Benefits Elicitation	Patient Forum	280 Benefits (at different levels)
23/09/2008			Benefits Criteria Workshop	6 Strategic Benefits Elicited (SOC)
08/12/2008			Benefits Elicitation	285 Sub & End Benefits Elicited
18/12/2008		Phase2: Benefits Classification	OPTIONING: 1st Workshop	Sub & End Benefits grouped in 30 sub-benefits – 7 designs down to 5
			Programme Board Meeting	Sign off of process to date
22/12/2008	Design Development 1:200	Phase2: Benefits Weighting	Programme Board Meeting	Look forward
23/01/2009			Programme Board Meeting	Selection of 1 out of 5 design options
09/02/2009			WEIGHTING: Benefits 2nd Scoring workshop	Benefits Weighting

Figure 12.6   BSUH 3Ts timeline summary.

Source: Based on Codinhoto and Passman (2009).

- description of benefits (or dis-benefits);
- identify risks and dependencies (e.g. among benefits and with other programmes/projects);
- discuss when benefits are expected to occur and over what period realisation will (most probably) take place;
- characterise the measure for valuating realisation of benefits (and how should be the measure carried out); estimate costs associated with benefits realisation and measurement;
- details of related changes to operations and staffing;
- distribute benefits responsibilities (e.g. assuring realisation) among trust division/clinical areas;
- identify to whom the benefit will mostly accrue: patients, visitors/carers, staff etc.

Sub and end benefits elicitation – Further *sub* and *end benefits* were elicited through 10 benefits identification workshops with groups of 20/25 BSUH 3T stakeholders[8] and researchers (facilitators). These workshops were used to gain the views of the different groups on BSUH 3Ts and the benefits they believe the BSUH 3T programme should bring. The workshops were very similar to a focus group, as they give stakeholders the chance to discuss, disagree and/or develop a shared perspective (Hakim, 2000).

---

[8] Stakeholder groups: Imaging, Cancer, HIV/Clinical Infection Service, Outpatients, Medicine/Elderly Care, Trauma/Critical Care, Neurosciences, Programme Board, Patient Representatives and Patient Experience Panel.

During the workshops an overall of 682 benefits were elicited. The benefits were then summarised and compiled into two main categories consisting of 8 *strategic benefits* and 37 *sub benefits* (presented in Table 12.5).

The benefits listed in Table 12.5 were then used to inform the optioning and weighting workshops, as explained in the following paragraphs.

Optioning: 1st Scoring Workshop (long list of design options) – At this stage five options plus two 'Do Minimum' design options, all 1:500 scale, were considered for BSUH 3Ts. The aim of this workshop was to get the proposed five design options down to three. It was decided that the 'Do minimum' design option had to go through to the next stage, as the baseline option to score the others against. A mix of 50/55 stakeholder[9] representatives were involved in this activity. Each design option was presented by the architect, and the group was asked to score each of these design options against the 8 strategic benefits and 37 sub benefits; scoring was 1–5, with 1 being low (does not meet criteria), 5 very good (should be taken forward). The exercise developed during this workshop resulted in three options being selected to move onto the next activity: weighting exercise, where the final selected option should be identified.

Weighting: 2nd Scoring Workshop (short list of design options) – The aim of this workshop was to agree on the preferred design option. It was facilitated by the BSUH 3Ts Project Director and a mix of stakeholders. The five design options were presented in detail to the group by the architect, the attendees were asked to assign a score of 1–5 against the different benefits for each design option. Once the scoring was completed the group undertook open discussions to rank the eight *strategic benefits*. This was done by getting to a general consensus of the order of importance of the benefits and assigning a percentage to each strategic and sub benefit, so that they totalled 100%. At the strategic (and sub) benefits level these benefits were weighted as detailed in Table 12.5.

Once these weightings were assigned to the strategic benefits, the same exercise and process occurred to assign weightings (%) to the sub benefits. It was then possible to score (between 1 and 5) the different options in relation to the ranking of the benefits and identify the final design proposal.

The 3Ts hospital development is now (May 2009) at a stage where the identified benefits can be fully incorporated in the comprehensive business case. These benefits details and the business case contents will further assist in identifying monitoring methods that need to be in place to support review and evaluation of benefits during the BeReal's *Phase 4: Benefits Evaluation and Reviews*. Benefits will be reviewed as necessary and evaluated as phase 4 materialises, since these will be used to demonstrate to what extent elicited (or new) strategic and sub benefits are being totally, partly or not realised.

---

[9]Stakeholders: 8–10 management, 4–5 patient group representatives, 5 commissioners, 3 local authorities, other (mainly, doctors and nurses).

12.5.2 Manchester, Salford and Trafford local improvement finance trust case study characterisation and discussion

In 2001, MaST LIFT was introduced as incorporating a new procurement route for primary care services through Community Health Partnerships (CHP). LIFTs intent is to contribute to the redevelopment of primary care infrastructure, through building facilities that can deliver diverse services including those for the acute sector (Binley, 2008). Currently, there are 42 nationwide LIFT schemes and this is set to continue in the future with 7 already under development, for which procurement activities are already ongoing.

MaST is the largest of the LIFT partnerships, and was established in March 2001. The HaCIRIC research team in collaboration with the MaST Partnership Programme Director selected 3 buildings from 12 operational schemes for the research, that is: (1) Partington Health and Social Care Centre, (2) Wythenshawe Forum Centre and (3) Douglas Green Energise Centre. The aim of the study was to evaluate the schemes at a post-occupancy phase in terms of benefits realised so far (planned vs. emerged) and to validate suitable methods for POE benefits measurement. Activities to achieve these aims include:

- The *creation* of a *steering group* for the project to ensure representation of key stakeholders, including, the MaST LIFT Partnership Programme Director, MaST LIFT Chief Executive, Manchester PCT Finance Director, the three Health Centre Managers, a DoH Gateway reviewer, a Manchester City Council Health Joint Unit program manager, a Primary Plus Facilities Manager, a Community Health Action Partnership Director and the research team. The group has scheduled quarterly meetings to review and monitor progress and give advice on future steps.
- Identification *of benefits related to 1st wave schemes of MaST LIFT*. This was a retrospective identification of benefits as the three schemes were already occupied and operational. In order to compile a catalogue of benefits to be evaluated, an initial study by the project team looked into the Strategic Service Development Plan (SSDP), the Local Development Plan (LDP) and the approved business case documents of the schemes. The result of the study delivered a first set of high-level benefits that the local healthcare authorities aimed to deliver through LIFT in the area of MaST. This was then further explored as part of a benefits identification workshop involving the Strategic Partnering Board (SPB) of MaST LIFT and the steering group. Based on that, the benefits identification workshop delivered a (second/reviewed) full set of benefits, organised into *strategic* and *sub* benefits. The set of benefits consist of over 5 *strategic* benefits and 36 *sub* benefits (see Table 12.6).
- Development and implementation of appropriate *POE*. Having a set of benefits to be evaluated, the project team moved onto identifying suitable methods for collecting data linked to these benefits in order to be able to evaluate to what extent these have been realised. Following a literature review on POE

Table 12.6 MaST LIFT Case Study – examples and benefits organisation.

Benefits	Benefits classification	
	Strategic benefits	Sub benefits
1. Improved patient services	✓	
1.01 Improved patient experience		✓
1.02 Better access to facilities (product)		✓
1.03 Greater privacy		✓
1.04 More services in 1 place (co-location)		✓
1.05 Improved health outcomes		✓
1.06 Greater access (service)		✓
1.07 Less waiting		
1.08 New services		✓
1.09 Care closer to home		✓
1.10 Increased patient choice		✓
2. Time cost quality	✓	
2.01 Faster procurement		✓
2.02 Faster delivery from concept to operation		✓
2.03 Removal of backlog maintenance		✓
2.04 Non-interruption of service product		✓
2.05 Predictability of time cost delivery		✓
2.06 Actual time cost delivery		✓
2.07 Flexibility and future proofing		✓
2.08 Cost savings due to co-location		✓
2.09 Lower total running costs		✓
3. Contribution to regeneration	✓	
3.01 Investment into deprived areas		✓
3.02 Higher local employment		✓
3.03 Improved community facilities		✓
3.04 Improved economic activity		✓
3.05 Sustainable environment (economic)		✓
3.06 Sustainable environment (social)		✓
3.07 Better links with other services – 'cause and effect'		✓
4. Improved staff satisfaction	✓	
4.01 Better working environment		✓

(Continued)

Table 12.6 (Continued).

	Benefits classification	
Benefits	Strategic benefits	Sub benefits
4.02 Incentives		✓
4.03 Reduced absences		✓
4.04 Increased career prospects		
4.05 Increased training opportunities		
4.06 Higher level of staff retention and increased corporate learning and memory		
5. Better partnership/continuous improvement	✓	
5.01 People working together on many schemes (greater understanding, reduced cost and time; better relationships, less conflict management)		✓
5.02 Increased quality between schemes		✓
5.03 Value for money improvement from scheme to scheme		✓
5.04 Access to finance		✓

Legend: ✓, Main focus.

and best practice evaluation guides[10] for healthcare facilities, the project team, with the steering committee, assigned a data collection method to each identified benefit. A combination of methods was identified, and Table 12.7 presents an extract of the benefits/methods matrix. The focus of the evaluation was mainly on capturing the perspective of service providers, facilities users[11] and of the community.

- *Validate a benefits mapping approach* and generate maps to include strategic, sub and end benefits, enablers and changes. A series of four workshops were organised aimed to test mapping techniques and produce a benefit dependency map based on the benefit set. The initial benefits mapping workshop involved members of MaST LIFT's SPB and was facilitated by the HaCIRIC research team. During the workshop, participants identified relationships between *strategic* and *sub* benefits using 'cause and effect' diagrams. There were two sub-

---

[10] For example, Association of University Directors of Estates (AUDE), Commission for Architecture and the Built Environment (CABE) and other best practice guides used for developing an effective questionnaire involved in POE, such as the CIM (DoH).

[11] *Patients, Centre Users* and all medical and administrative *staff* working in the LIFT centres including GP Practice, Pharmacy, Dentistry, Physiotherapy, Orthoptics, Audiology, District Nursing, Health Education, Speech Therapy, Anti-coagulation, Community Health Action Partnership, Citizens Advice and Community services.

Table 12.7   Methods of 'actual (emergent)' data collection – MaST LIFT methods and benefits matrix.

| | Methods of data collection | | | | |
| | Questionnaires[a] | | | | Data sources |
Benefits	Patient	Community	Staff	Interview[b]	(existing)
1. Improved patient services					
1.01 Improved patient experience[c]	✓				
1.02 Better access to facilities (product)	✓	✓	✓	✓	
1.03 Greater privacy	✓		✓		
1.04 More services in one place (co-location)		✓			✓
1.05 Improved health outcomes					✓
1.06 Greater access (service)	✓	✓		✓	✓
1.07 Less waiting	✓				✓
1.08 New services	✓	✓			✓
1.09 Care closer to home					✓
1.10 Increased patient choice	✓	✓			✓

Legend: ✓, Main focus.
References: Sapountzis et al. (2008) and Yates et al. (2009).
[a]*Questionnaires* provide a relatively quick and low-cost way to gain a reliable insight into characteristics, attitudes and beliefs' of a large sample (Marshall and Rossman, 1999; May, 1993).
[b]*Interviewing* is the most significant qualitative technique that can be used within a case study to obtain information from a small sample (Easterby-Smith et al., 2002; Yin, 2003). For this case study, semi-structured interviews are used to collect qualitative data. Semi-structured interviews have some predetermined questions, but these are used only as a guide, the researcher is able to change these and explore further into an answer, dependant on the response and how the researcher interprets the conversation (May, 1993; Robson, 2003).
[c]Further details in terms of the questionnaire survey approach is given in Table 12.8.

sequent workshops with the steering group to further enhance and validate the map. The logic of the map was to work from right to left when linking *strategic* to *sub* benefits and from left to right when linking *enablers* or *changes* to benefits. An illustrative example is shown in Figure 12.7.

- Evaluation of the perceived impact according to patients, staff and centre users of the three evaluated schemes in relation to the five MaST LIFT strategic benefits. The research team used questionnaires for staff, patients and centre users, and interviews to produce primary quantitative and qualitative data. A 35% sample size of the patients' total number was targeted. (The percentage includes an estimated 40% non-response rate.) A 15% return on patient centre/user

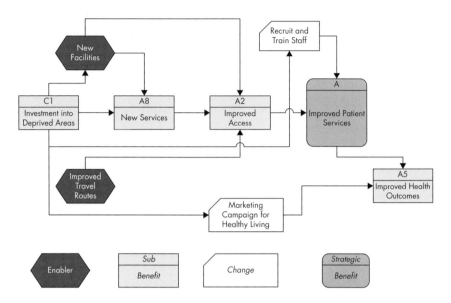

Figure 12.7   Benefits map: MaST LIFT strategic benefit '1. Improved Patient Services'.
Source: MaST LIFT Case Study data (adapted); content is illustrative.

questionnaires distributed was achieved. The low number of staff working in the centre in each working group/discipline indicated that the whole working population was targeted. A 70% return on staff questionnaire distributed was achieved.

The aim of the questionnaire was to link each benefit that was going to be evaluated to a specific question. Therefore, the research team would be able to analyse to what extent patients, staff and centre users perceive that the benefit has been realised (totally, partially). Table 12.8 gives an illustration of some questions included in the first section of the Patient/Centre User questionnaire, as it was distributed in the pilot run of the questionnaire using samples from the Douglas Green Energise Centre.

Analytical software like NUD*IST[12] and SPSS[13] were used to analyse both the quantitative and qualitative data as it was emerged from the questionnaires. Table 12.9 shows how the questions and the survey responses were linked to specific benefits, using cumulative satisfaction level of the data analysed.

---

[12] Non-numerical, Unstructured Data Indexing, Searching and Theorising (NUD*IST) software is used to analyse the qualitative data. This software is beneficial when looking into large descriptive texts (Yin, 2003) that may have resulted from the semi-structured interview as it is based on a code-and-retrieve facility. This facility codes large amounts of data, which can be later on retrieved through using 'Boolean, context, proximity, and sequencing searches' to identify qualitative matrices, relationships and patterns (Richards and Richards, 1994).

[13] Statistical Product and Service Solutions (SPSS) – This is a suite of programmes which form a computer package used by business organisations, education and health researchers, survey companies, the government and others. Through statistical analysis of data it can carry out in-depth data access and preparation, analytical reporting and graphics, as well as major statistical aspects.

Table 12.8   Questionnaire used for 'Actual (emergent)' data collection MaST LIFT pilot – Section 1 for the patient/centre user.

Benefits	Benefits classification	
	Strategic benefits	Sub/end benefits
1. How satisfied are you with	✓	
1.1 Accessibility of the building from the street		✓
1.2 The visibility of the building's entrance		✓
1.3 Signs to guide patients and visitors to the centre		✓
1.4 Ability to get to the centre by car		✓
1.5 Ability to get to the centre by public transport		✓
1.6 Peace and quiet of the centre		✓
1.7 Fresh air and ventilation within the centre		
1.8 Space around the examination/operation couch		✓
1.9 Location and availability of toilets		✓
. . .	. . .	. . .

Legend: ✓, Main focus.

Source: MaST LIFT Case Study data; adapted/content is not exhaustive.

Table 12.9   Example of cumulative satisfaction levels – correlation of questions with benefits.

Individual benefits[a]	Results (%)[b]	Questionnaire answers (%)[b]						
1.1 *Improved patient experience*		1.6	1.7	1.9	1.11	1.12	1.13	. . .
Very disatisfied	3.2	2.4	2.4	2.6	2.4	3.3	6.9	. . .
Dissatisfied	1.9	0.0	4.9	0.0	0.0	0.0	3.4	. . .
Fairly satisfied	13.9	2.4	4.9	5.1	7.3	6.7	17.2	. . .
Satisfied	32.8	26.8	31.7	28.2	34.1	46.7	41.4	. . .
Very satisfied	48.2	68.3	56.1	64.1	56.1	43.3	31.0	. . .

12.6 Conclusions

Benefits realisation is emerging as a vital element for 'best practice' programme management in the context of healthcare capital investment in the UK. This chapter presented findings from a literature review in support to that statement, introduced an emergent high-level BeReal and highlighted findings from two case studies, targeting the integrated planning and evaluation of healthcare infrastructures and services in the UK. The research presented provides evidence of the importance of benefits realisation along different phases of capital investment programmes and the findings from the case studies have contributed to the development and validation of the major different dimensions within the BeReal model, as discussed along the following paragraphs.

Benefit taxonomy – Development of benefits taxonomy has been assured, under a dual approach of classification and characterisation. Detailed specification and benefits is not considered before the development of the full business case (informed by *BeReal phase 2*). Complementarily to the benefits organisation vision, is now more clear that benefits might be realised at different phases of a programme's life cycle, although major impacts will occur in later phases of the built environment life cycle; nevertheless, benefits realisation mostly occur during and after the construction phase.

The organisation of benefits in terms of a selected group of aspects (e.g. planned and unplanned, internal and external, built environment life cycle states, stakeholder category) using segmentation techniques, is regarded as advantageous, especially to understand and clarify which aspects might be more (or less) represented by the elicited group of benefits. It also helps to understand benefit ownership issues and when a benefit should be measured and realised all of which contribute to the monitoring and managing aspects of the tool.

Techniques and procedures for data collection and data analysis – Active and systematic organisation of the identified/elicited benefits, under a three-level organisation structure, is regarded as a necessary and valuable activity. This activity consists of highlighting (dis)similarities, consolidating (e.g. two similar elicited benefits in one) and segmenting elicited data/benefits under the systematised dual organisation approach that is able to assure support throughout the investment programme (e.g. selection of design options, controlling/monitoring).

Since benefits are elicited a proper traceability management of benefits is recommended, highlighting stakeholders involved, identified overlapping and dependencies between the benefits etc. Major data collection techniques tested and recommended to elicited benefits are workshops with stakeholders, surveying questionnaires and historical data gathering through consultation of existing documentation.

Stakeholders – Benefits elicitation meetings with a diversity of stakeholders are recommended and understood as a critical surveying activity, since the participation of a variety of stakeholders enables the incorporation of different views and perspectives. Participation of a diversity of *stakeholders* (including the overall programme management team) along the programme life cycle and throughout the organisation (e.g. business functions) is also regarded as beneficial under a management of expectation perspective and contributes to a better comprehensiveness (scope) of benefits.

Collective decision-making – Should be developed under a sequential mode optioning (and weighting) approach. Firstly, selection of options (optioning) based on weighting strategic and sub benefits highly enables decision-making among a higher number of different design options, since it concentrates decision in a reduced number of benefits and does not include too much benefit's details (e.g. end benefits) into the decision-making process. Secondly, the identification of the best option (ranking) is based on a ranking developed only among the shortlisted options, and also focusing on the strategic and sub benefits levels. Only after ranking the shortlisted options, further detailed benefits are incorporated into the decision-making process as a basis for the development of a full business case (see *BeReal phase* 2) approach.

Monitoring/controlling structure – Conceptual design of a monitoring/controlling structure covering the overall built environment life cycle and organisational views, should guarantee traceability of elicitation/changes along the programme life cycle, highlighting dependency and overlapping of benefits. Based on conversations held with the programmes' management teams, the cross-analysis deviation between what benefits have been *planned* (to-be), what is now in place (*as-is or actual*) and what is in fact *delivered (emerged)* creates an effective life cycle monitoring/controlling approach of deviations that seems to be extremely relevant to properly measure the success of the programme, and that should cover both *efficiency* (resources versus results) and *effectiveness* (achievement of what is planned) perspectives.

12.7 Acknowledgements

This work was funded by the Engineering and Physical Sciences Research Council and conducted through the Salford Centre for Research and Innovation and the Health and Care Infrastructure Research and Innovation Centre.

References

Ashurst, C. and Doherty, N. F. (2003) Towards the formulation of 'a best practice' framework for benefits realisation in IT projects. *Electronic Journal of Information Systems Evaluation*, 6: 1–10.

Bartlett, J. (2006) *Managing programmes of business change*, 4th edn., Project Manager Today, Hampshire, UK.

Bennington, P. and Baccarini, D. (2004) Project benefits management in IT projects – An Australian perspective. *Project Management Journal*, 35(2): 20–30.

Binley (2008) *NHS Guide*, Summer–Autumn edn. Beechwood House Publishing, Basildon.

Bradley, G. (2006) *Benefit realisation management – A practical guide to achieving benefits through change*. Gower, Hampshire, UK.

Brighton and Sussex University Hospitals. (2009) Available at: http://www.bsuh.nhs.uk/about-us/3t/, accessed February 23rd, 2009.

BSUH 3Ts documentation. (2008) *3T Service Modernisation – Update*, 13th January, 2009.

Carruthers, J., Ashill, N. J. and Rod, M. (2006) Mapping and assessing the key management issues influencing UK public healthcare purchaser-provider cooperation. *Qualitative Market Research: An International Journal*, 9(1): 86–102.

CCTA (Central Computer and Telecommunications Agency). (1999) *Managing successful programmes*. The Stationery Office, London.

CCTA (Central Computer and Telecommunications Agency). (2000) Managing the business benefits. *The Antidote*, 27: 28–29.

Codinhoto, R. and Passman, D. (2009) *3Ts Sussex Hospital – Optioning meeting*, internal process report, University of Salford 9th February 2009.

Cooke-Davies, T. (2002) The real success factors on projects. *International Journal of Project Management*, 20: 185–190.

DoH website. (2008) Available at: http://www.dh.gov.uk/en/AdvanceSearchResult/index.htm?searchTerms=patient%20and%20public%20involvement%20forums.

Easterby-Smith, M., Thorpe, R. and Lowe, A. (2002) *Management research: An introduction to research methods*. Sage, London.

Farbey, B., Land, F. and Targett, D. (1999) The moving staircase – Problems of appraisal and evaluation in a turbulent environment. *Information Technology and People Journal*, 12: 238–252.

Flyvbjerg, B. (2007) Five misunderstandings about case study research. *Quality Inquiry*, 12(2): 219–245, Sage Publications.

Glynne, P. (2007) Benefits management – Changing the focus of delivery. *Association for Progress Management Yearbook 2006/07*, pp. 45–49.

Hakim, C. (2000) *Research design: Successful designs for social and economic research*, 2nd edn. Routledge, London.

Herriot, R. E. and Firestone, W. A. (1983) Multisite qualitative policy research: Optimizing description and generalizability. *Educational Researcher*, 12(3): 14–19.

Holton, E. F. III and Burnett, M. F. (1997) *Quantitative research methods*. In Swanson, R. A. and Holton, E. F. III (eds), *Human resource development research handbook: Linking research and practice* (pp. 65–87). Berrett-Koehler, San Francisco, CA.

Kagioglou, M., Cooper, R., Aouad, G. and Sexton, M. (2000) Rethinking construction: The Generic Design and Construction Process Protocol. *Journal of Engineering Construction and Architectural Management*, 7(2): 141–154.

Lewin, K. (1946) Action research and minority problems. *Journal of Social Issues*, 2(4): 34–46.

Mantzana, V. and Themistocleous, M. (2004) Identifying and classifying benefits of integrated healthcare systems using an actor-oriented approach. *Journal of Computing and Information Technology (CIT)*, 12: 265–278.

Marshall, C. and Rossman, G. B. (1999) *Designing qualitative research*, 3rd edn. Sage, Thousand Oaks, CA.

May, T. (1993) *Social research: Issues, methods and process*. Open University Press, Buckingham.

McCartney, R. (2000) Successful IT: Modernising government in action. Report, Cabinet Office UK, p. 35.

NHS. (2007) No delays: No delays archive, available at: http://www.nhs.yk/serviceimprovement/tools/ITOII-Benefitsrealisation, accessed August 8th, 2008.

Nogeste, K. (2006) *Development of a method to improve the definition and alignment of intangible project outcomes with tangible project outputs*. Graduate School of Business, RMIT University, Melbourne, p. 339.

Nogeste, K. and Walker, D. H. T. (2005) Project outcomes and outputs: Making the intangible tangible. *Measuring Business Excellence*, 9: 55–68.

OGC. (2007a) *Managing successful programmes*. The Stationery Office, London.

OGC. (2007b) STDK home, delivery lifecycle: Benefits management, available at: http://www.ogc.gov.uk/sdtoolkit/reference/deliverylifecycle/benefits_mgmt.html#benmanagement1, accessed February 7th, 2009.

Punch, K. F. (2005) *Introduction to social research: Quantitative and qualitative approaches*, 2nd edn. Sage, Thousand Oaks, CA.

Reiss, G., Anthony, M., Chapman, J., Leigh, G., Pyne, A. and Rayner, P. (2006) *Gower handbook of programme management*. Gower Publishing, Aldershot.

Remenyi, D. and Sherwood-Smith, M. (1998) Business benefits from information systems through an active benefits realisation programme. *International Journal of Project Management*, 16: 81–98.

Richards, T. J. and Richards, L. (1994) Using computers in qualitative research. In Denzin, N. K. and Lincoln Y. S. (eds), *Handbook of qualitative research*. Sage Publications, Thousand Oaks, CA.

Robson, C. (2003) *Real world research*, 2nd edn. Blackwell Publishing, Oxford.

Sakar, P. and Widestadh, C. (2005) *Benefits management – How to realize the benefits of IS/IT investments*. Department of Informatics, University of Goteborg, Goteborg.

Sapountzis, S., Harris, K. and Kagioglou, M. (2008) The need for benefits realisation – Creating a benefits driven culture in UK's Healthcare Sector. *1st HaCIRIC Symposium 'Redefining healthcare infrastructure Integrating services, technologies and the built environment'*, London, UK.

Sapountzis, S., Yates, K., Kagioglou, M. and Aouad, G., (2009) Realising benefits for primary healthcare infrastructures. *Facilities Journal*, 27(3/4): 74–78.

Silva, F. V. (1995) *Contabilidade industrial*, 9th edn. Livraria Sá da Costa Editora, Lisbon.

Silverman, D. (2006) *Interpreting qualitative data: Methods for analyzing talk, text and interaction*, 3rd edn. Sage Publications, London.

Simon, T. (2003) What is benefits realisation? *The Public Manager*, Winter 2003–2004, 59–60. In Nogeste, K. (2006) *Development of a method to improve the definition and alignment of intangible project outcomes with tangible project outputs*. School of Business, RMIT, Melbourne.

Sonali, K. S. and Corley, K. G. (2006) Building better theory by bridging the quantitative–qualitative divide. *Journal of Management Studies*, 43(8): 1822–1835.

Sweeney, K. and Griffiths, F. (2002) *Complexity and health care: An introduction*. Radcliffe Medical Press, Oxford.

Thorp, J. (1998) *The information paradox – Realising the business benefits of information technology*. McGraw-Hill, Toronto, Canada.

van Strien, P. J. (1975) Naar een methodologie van het praktijk denken in de sociale wen in de Sociale Wentenschappen. *Naderlands Tidschrift voor de Psychologie*, 30: 60–619. In Fellows, R. and Liu, A. (eds) (2003), *Research methods for construction*, 2nd edn. Blackwell, Oxford.

Ward, J. and Daniel, E. (2006) *Benefits management – Delivering value from IS & IT investments*. Wiley, West Sussex, UK.

Ward, J. and Elvin, R. (1999) A new framework for managing IT-enabled business change. *Information Systems Journal*, 9: 197–222.

Ward, J., Murray, P. and Daniel, D. E. (2004) *Benefits management best practice guidelines*. Cranfield University, Bedford.

Ward, J., Taylor, P. and Bond, P. (1995) Identification, realisation and measurement of IS/IT benefits: An empirical study of current practice. In Brown, A. and Remenyi, D. (eds), *Second European Conference on Information Technology Investment Evaluation*, Henley Management College, Henley on Thames.

Ward, J., Taylor, P. and Bond, P. (1996) Evaluation and realization of IS/IT benefits: An empirical study of current practice. *European Journal of Information Systems*, 4: 214–225.

Yates, K., Barreiro Lima, J., Sapountzis, S., Tzortzopoulos, P. and Kagioglou, M. (2009) BeReal benefits realisation model integrated approach: The built environment lifecycle and organisational views. *2nd HaCIRIC International Symposium*, April 2009, Brighton, UK.

Yin, R. (2003) *Case study research: design and methods*, 3rd edn. Sage Publications, Thousand Oaks, CA.

Achieving Continuous Improvement in the UK Local Improvement Finance Trust (LIFT) Initiative

13

A.D. Ibrahim, A.D.F. Price and A.R.J. Dainty

In the UK as well as in many other countries, the primary care sector emphasises the movement of healthcare out of large institutions, such as hospitals, into community-based settings, thereby bringing care closer to the people and making it more responsive to their needs (Baggot, 2004; Nwakoby, 2004). Until recently, investments targeted at providing facilities to support the provision of modern and integrated primary and social care in the UK have been fragmented and piecemeal (NAO, 2005). Consequently, the Local Improvement Finance Trust (LIFT) initiative was announced by the UK Department of Health (DoH) in 2000 as a way of mobilising huge investments to improve the quality of primary care buildings, particularly in deprived urban areas. The initiative aims to deliver a step change in the quality of the primary care estate, remedy some of the deficiencies in the existing arrangements and contribute to delivery of the investment targets identified within the *NHS Plan* (DoH, 2000). According to the NAO (2005), the objectives of the initiative include helping in:

- bringing significant improvements to GP premises;
- supporting co-location of healthcare professionals;

- forging links between primary and social care;
- supporting reduction in difficulties associated with GP recruitment and retention;
- shifting services away from the secondary care level;
- assisting in the achievement of good chronic disease management;
- enhancing 'Patient Choice' by providing patients with more choice over how, when and where they receive treatment.

The LIFT procurement philosophy embodies an integrative way of working between organisations from public and private sectors to deliver improvements in primary care estate (DoH, 2001). The execution of LIFT schemes involves intricate processes and complex interactions amongst and between large supply chains over a typical period of 20–25 years. Communication of vital knowledge and information between the different stages of these projects and across the disparate groups that are involved offer significant challenges in terms of efficiency, effectiveness and interface management (Ibrahim and Price, 2005a).

The LIFT Companies (*LIFTCos.*) are set-up as public–private partnerships (PPPs) in the form of limited liability companies under Strategic Partnering Agreements (SPAs) to deliver investment and services in primary care facilities within their localities. The SPAs used under the LIFT procurement explicitly require that schemes demonstrate the delivery of value for money (by market testing) both at commencement and at five yearly intervals. This demonstration of value for money is used as a fundamental approval criterion within the procurement process. Recent evaluations of the initiative have revealed mixed results. For example, on one hand the NAO (2005) report portrayed LIFT as an attractive way of securing improvements in primary and social care and that the schemes examined were effective and offered value for money. Hudson *et al.* (2003) used the NHS Environmental Assessment Toolkit (NEAT) to demonstrate that LIFT schemes are delivering more sustainable solutions compared to primary care facilities delivered through the traditional procurement routes. Ibrahim *et al.* (2008) showed that there is considerable evidence that the LIFT procurement is delivering the expected economies of scale in providing modern facilities for the provision of integrated primary and social care services. They also revealed that there were significant differences in the maturity levels of the schemes evaluated in terms of appropriate systems, processes and structures in the planning and implementation of the schemes. The pattern of progress made generally confirmed an evolving system, with some evidence of project-to-project performance improvement such as reduced tendering and legal costs, increased speed of completion as well as better quality facilities. Yet, it was established that there is potential for more improvements.

NAO (2005) also revealed that the attainment of the contractual requirements for both the demand and supply sides to continuously improve performance under the LIFT scheme remains unsystematic and may thus be unsustainable. Holmes *et al.* (2006) found from a case study involving two LIFT schemes that because the bidding processes typically involved unequal struggles between large consortia and inexperienced clients, the demonstration of value for money had been difficult and resulted in wasted opportunities in obtaining optimum designs and prices. Furthermore, although LIFT initiative advocates cross-project and scheme-wide learning, NAO (2005) revealed that there was little evidence of knowledge sharing between the PCTs they evaluated and that subsequent projects are already being embarked upon. These have highlighted the need for greater interest in evaluating how LIFT schemes are set-up and the arrangements for ongoing value for money and accountability assessments as well the demonstration of continuous improvement (CI).

The motivation for developing a continuous improvement framework (CIf) is hinged on the overwhelming evidence from previous research that the systematic harvesting of project experiences can enable organisations to develop project competences that lead to sustainable competitive advantage through the documentation of its most effective problem-solving mechanisms in a way that facilitate the logical demonstration of CI from project-to-project (Barlow and Jashapara, 1998; Chinowsky *et al.*, 2007; Schindler and Eppler, 2003; Tan *et al.*, 2007). In addition, the systematic documentation of mistakes, mishaps and pitfalls should also help an organisation to reduce the risks associated with future similar projects (Imai, 1986; Tan *et al.*, 2007). However, previous research works have focused primarily on harvesting explicit knowledge that are relatively easy to document (such as costs, timelines or other quantitative data) and mostly comprise numerical data that answers 'what', 'where' and 'how many' questions. Furthermore, because the numerical data often do not provide answers to other key pressing project questions and problems such as the reasons for failure or how particularly efficient solutions have been built or how certain special issues have been addressed, the consideration of tacit aspects to address questions such as 'know-how' (procedural or heuristic knowledge) and especially 'know-why' (such as experiences and insights into cause–effect relationships) are now being advocated (Williams *et al.*, 2001).

Accordingly, this chapter describes research whose objectives were to: explore the CI concept in general; identify the requirements for implementing CI in LIFT; develop a generic framework for achieving CI in LIFT procurement; and identify the key application challenges within the context of LIFT procurement initiative. For more details on the LIFT process, please see Chapter 5. The succeeding sections commence with a critical review of literature on CI concept and characteristics with the view to identifying the essential ingredients for achieving CI in construction environments. The methodology followed to explore the essential requirements for attaining CI in LIFT projects was outlined. The findings from data collected and analysed were presented to reveal the concept, implementation requirements for achieving CI in LIFT projects as well as the key application challenges associated with the LIFT procurement initiative. A conceptual CI model and a generic CI framework for LIFT were subsequently developed along with an operationalisation strategy.

13.2 Continuous improvement concept

During the past two decades, there has been growing interest in the concept of incremental innovation as a route towards improvement of various aspects of manufacturing (Bessant *et al.*, 1994; Steele and Murray, 2004) and, more recently, construction (Davey *et al.*, 2004; Maqsood *et al.*, 2007; Powell, 1999; Slaughter, 1998, 2000). This has resulted from increasing demand for accountability and performance expectations on organisations to continuously improve their efficiency and effectiveness. Consequently, the phrase 'continuous improvement' has become increasingly popular and has been associated with a variety of organisational development initiatives.

The CI philosophy adopts the stance that a development process is a continuum (Oakland, 1995) and that improvements only occur if attempts are made to learn from new information generated by the process itself rather than the product (Cooper *et al.*, 2005). The CI process is commonly associated with the *plan-do-check-act* (PDCA) cycle (Tague, 2004), with each phase of the cycle playing a very important role in sustaining improvement in an ongoing fashion. Suzaki (1987) defined CI as '*incremental improvement of products, processes, or services over time, with the goal of reducing waste to improve workplace functionality, customer service, or product performance*'. Bessant *et al.* (1994) defined CI as '*a company-wide process of focussed and continuous incremental innovation sustained over a long period of time*'. Juergensen (2000) defined CI as consisting of '*improvement initiatives that increase successes and reduce failures*'. It is clear from these definitions that CI is multi-faceted and extends across several organisational development initiatives. Although the concept is clear, the diversity in implementation subjects it to varied understandings and interpretations. Nonetheless, it is clear that the concept is concerned with an effort to upgrade the performance of every facet of an organisation and covers more than simply improving the quality of the (built) products. However, a common understanding and explicit tools and techniques are essential for any meaningful progress in enshrining it in the culture and practices of the construction industry.

Available literature on CI shows that it has no universal theoretical basis (Savolainen, 1998) as it tends to be used as a generic term that has acquired many of its attributes from other initiatives including total quality management (TQM), just-in-time (JIT) production system, lean techniques, six sigma, employee involvement programmes etc. However, there is a growing recognition of its application in other areas such as cost reduction, value enhancement, flexibility, waste minimisation, inter-organisational relations and for supporting process improvement (Caffyn, 1999; Gallagher *et al.*, 1997; Imai, 1986; Oakland, 1999; Robinson, 1991). Clearly, these goals are applicable to many facets of the construction process and have contributed to the growing interests in managerial concepts such as partnering, lean construction, sustainable construction, value management and TQM in the construction sector.

Hill (1996) argued that CI and learning are inextricably linked such that learning is the most compelling reason for undertaking any CI within an organisation. Whetherill *et al.* (2002) also asserted that an organisation's only sustainable advantage lies in its capability to learn faster than its competitors and the rate of change imposed by the external environment, and that there is a need to 'integrate learning within day-to-day work processes'. In a study of strategic change in four UK industries, Pettigrew and Whipp (1991) concluded that more successful firms had developed effective learning processes at all levels of their organisation. To remain successful though, Barlow and Jashapara (1998) argued that learning needs to be dynamic and evolving along with the competitive forces at play. This will necessitate a shift from an essentially static approach to learning based on information acquisition towards a greater emphasis on information interpretation, distribution and adaptation.

According to Barlow and Jashapara (1998), the nature of organisational learning in a particular industry is dependent to a large extent on factors such as the dominant competitive environment and the size and underlying cultural assumptions and values of organisations in the industry. While considerable success has been recorded in the manufacturing and automotive industries through the application of CI methodologies (Anumba *et al.*, 2000; Errasti *et al.*, 2007; Green *et al.*, 2004), the construction industry is still lagging behind in adopting these technologies and management techniques for performance improvement (Ferng and Price, 2005). The highly fragmented nature of the market and structure of the construction industry as well as over-emphasis on contractor selection based on lowest price often result in organisations being locked up together with overly restricted forms of contract, resulting in high levels of claims, counter-claims, litigation and dissatisfied clients (Egan, 1998; Latham, 1994). The dichotomy between and amongst the demand and supply sides of construction projects has consequently made 'learning' about the market's changing needs difficult (Thomas and Thomas, 2005). The unique and transient nature of many construction projects also deters any attempt to standardise the process-steps, consequently resulting in lost opportunities in feeding back the experiences gained on projects for the benefits of future projects (Egan, 1998; Latham, 1994).

These historical and cultural factors associated with the construction industry coupled with the deep fragmentations in the healthcare sectors along professional divides (Grimsey and Graham, 1997) militate against a more dynamic approach to learning in the procurement of healthcare facilities in general. Moreover, the call for the construction industry to look at other sectors for best practices has attracted varied responses from the academia and industry. On one hand, researchers and practitioners have searched for related good practices that have been successfully adopted and implemented in other industries (e.g. Anumba *et al.*, 2000; Errasti *et al.*, 2007; Fernie *et al.*, 2001, 2002, 2003; Green *et al.*, 2002, 2004; Ngowi, 2000), principally from manufacturing and to a lesser degree the service sectors. The underlying assumption being that borrowing something that has gained acceptance in other industries, rather than inventing a new solution, is easier to exploit (Towill, 2003). On the other hand, practices originating from other industries or other construction projects have been rejected on the basis of being inappropriate because the characteristics of construction and of each project are perceived as 'unique' both in terms of discontinuities and the fragmentation of the teams into different professional disciplines (Ahmad and Sein, 1997; Bresnen, 1990; Bresnen *et al.*, 2003; Pasquire and Connolly, 2002). While these two contrasting views may not be necessarily mutually exclusive, this research has adopted Lillrank's (1995) suggestion that good practice adopted elsewhere can be exploited, provided that it is sufficiently adapted to the new situation.

Consequently, improving the performance and competitiveness of construction projects in the healthcare sector would require both cultural and behavioural shifts in the mindsets of practitioners and the various professional groups. These shifts will require the development of new skills and techniques and the 'unlearning' of traditional practices or, as Kumaraswamy (1998) noted, the structural arrangements may need to be 'reconstructed' so that constant innovation and CI can be encouraged to become the norm. The importance of organisational values to the success of CI is also widely recognised, but these have been usually addressed only within other contexts such as TQM or partnering. The behaviour and culture of the people in an organisation have been adjudged to be a reflection of the organisation's values and priorities (Berger, 1997; Bessant and Caffyn, 1997; Hyland *et al.*, 2000; Jabnoun, 2001). Garfield (1986) contended that commitment to values is a key determining factor in the pursuit of any mission. Consequently, the identification of values that motivate and sustain CI became pertinent. In taking a long-term strategic approach, Bessant *et al.* (1994) also argued that success and sustainability of the CI initiatives will depend on the exploration of *driving* and *enabling* values. While there are some general prescriptions for creating a suitable environment for establishing an innovative culture, there is little systematic research on the specific requirements for CI, or how these might vary according to different organisations (Bessant *et al.*, 1994; Errasti *et al.*, 2007; Kerrin, 1999; Tennant *et al.*, 2002). This is particularly important for LIFT procurement which involves long-term partnerships and associated opportunities for continuous performance improvement. Although LIFT is 'local' by nature and operates under varying social and economic landscapes thereby creating difficulty in applying lessons learnt elsewhere without appropriate contextualisation, construction projects are largely similar at micro level in terms of processes and resources. Consequently, some of the lessons and knowledge generated during their execution can be reused in future phases and projects. The methodology followed to explore the essential requirements for achieving continuous performance improvement in LIFT projects is described in the following section.

13.3 Research method

This exploratory research adopted a qualitative approach in order to gain detailed insight into current practices. Because the researchers had no *a priori* knowledge of the actual practices adopted by the organisations from which a hypothesis could be derived, an inductive methodology was chosen. More so, given the diversity of the participants involved in each LIFT scheme, it was necessary to explore the perspectives of the principal stakeholders.

Data were collected from secondary and primary sources: project archival research and interviews with stakeholders. Project archives held by the clients, consultants, the private sector partners and other public sector partners were examined. Semi-structured interviews were conducted across a broad constituency of stakeholders, comprising ten senior managers from six organisations working on three LIFT schemes (with one each representing first, second and third wave[1] of LIFT schemes). Interviewees were drawn from the following categories:

- public sector central lead Project Directors;
- public sector Directors of Primary Care representing the PCTs;
- public sector Project Managers;
- Chair and General Managers of the LIFTCos. (from private sector);
- public sector independent technical consultants (from private sector).

Six of the interviewees were from the public sector while the remaining four were from the private sector. Each interview lasted between 45 min and 2 h. The semi-structured nature of the interview technique allowed the interviewees to elaborate on any topic, but required all predetermined topics to be covered. The diversity in the interviewees enabled a broad cross-section of views to be canvassed in relation to the efficacy of the LIFT procurement approach. To enable the development of CI framework for LIFT procurement, the interview questions explored CI concept in general; the requirements for implementing CI in LIFT; and the key application challenges associated with the LIFT procurement initiative. In addition, supporting documentation was provided and used as supplementary information which was analysed and evaluated.

The interviews were all recorded, transcribed and analysed using the constant comparative analysis technique. The constant comparative analysis technique shares much with the Grounded Theory methodology, the basic difference being that the former is exploratory in nature and does not extend to the establishment of a theory from the data collected. The technique involves taking one piece of

---

[1] By June 2004, 42 LIFT schemes had been go ahead in three waves (xx in first wave, xx in second wave and xx in the third wave). By November 2004, seven fourth wave LIFT schemes were announced by the Secretary of State for Health. All the schemes under the first three waves have reached financial close, and several are proceeding towards second and subsequent financial closes. However, the procurement process for the fourth wave schemes is under way with a number of projects having selected their preferred partners.

data (e.g. one interview, statement or theme) and comparing it with all others that may be similar or different in order to develop conceptualisations of the possible relations between various pieces of data. For example, by comparing the accounts of two different people who had a similar experience, a researcher can pose analytical questions around the differences between responses and how categories relate. While the authors have sought to be fair and accurate in conveying the informants' responses, it should be noted that the sample of schemes considered is relatively small and thus, the findings may not be necessarily be reflective of the entire LIFT programme at large.

13.4 Result and discussions

The results of the data analyses (both from interviews and supporting documentations) are presented in the following paragraphs.

The interviews revealed that the participants in LIFT procurement have some knowledge of the requirements of CI, but that current practices are ad hoc, unstructured and unsystematic. They also revealed that many of the organisations are aware of the potential competitive advantage obtainable from harvesting individual experiences (tacit knowledge) into well-structured explicit knowledge to be reused, and different technological tools such as applications based on advanced databases, the Internet and groupware technologies have been developed to support these transmission processes. However, most of the existing tools only process data and information and often ignore the personal dimensions associated with the process. For example, the existing tools often underscore the need to stimulate individual affection in order to generate knowledge as knowledge cannot be extracted from individuals without their participation and motivation.

It was further established that it has become essential to develop a comprehensive framework that considers the human values and ensures that stimulation of learning during the planning, design, construction, occupancy and post occupancy become a natural part of the process of procuring primary care facilities. More specifically, the framework should provide a simplified model of the complex relationships amongst the factors that influence the primary care sector (both from within and outside the sector) in the achievement of CI in the procurement process.

13.4.1 CI concept

The research revealed that CI is interpreted differently by different organisations and project participants. Some of the phrases/sentences used to describe CI by the interviewees include:

- quality improvement;
- application of innovative technologies and techniques to enhance functionality of a project;

- application of innovation to improve project outcomes in terms of time, cost and quality;
- improved satisfaction for all stakeholders through value-enhancement mechanisms;
- improved efficiency to deliver projects at lower cost to the client;
- demonstration of learning from project-to-project in terms of project outcomes of time, cost and quality;
- improved effectiveness to deliver projects faster to the client;
- improved client satisfaction with the overall project.

Across the three schemes, although the interviewees recognised the importance of learning from project-to-project and from other schemes, there were no formal structures to facilitate effective knowledge capturing or sharing in the format that it can be effectively reused in subsequent phases or projects. It was established that 'lessons learnt reviews' were carried out at both strategic and operational levels in one of the schemes, but the reviews were restricted to the commissioning process only. The three schemes have been generally more inward-looking, relying on reflective and audit trailing techniques in their governance departments. However, the interviewees recognised the need for broader perspectives through the use of more systematic and structured approaches for learning from within and outside their schemes.

13.4.2 Essential requirements of continuous improvement in LIFT
13.4.2.1 Preconditions and success factors for CI
In order to enshrine CI culture in the employees and hence the organisation, the following preconditions were identified by the interviewees:

- establishment of common understanding in what CI is; what the targets are; what the success criteria are; and what the roles, responsibilities, accountabilities and lines of communications are at individual, team and organisational levels;
- availability and/or access to appropriate competence/skills for CI, and these were identified as systematic problem solving, experimentation with new approaches, learning from own experience and past history, learning from experience and best practices of others and transferring knowledge quickly and efficiently throughout the organisation;
- joint creation of supporting/enabling processes that are understood by all;
- commitment and trust at all levels of each participating organisation.

13.4.2.2 CI driving values
In projects involving multiple participants from diverse sectors such as LIFT projects, the driving values, which provide answers to the questions 'why should we continuously improve' and 'what values should we have to want to improve'

were recognised by the interviewees as the basic values necessary for establishing and reinforcing commitment to CI. They were identified to include:

- mutual respect, both vertically through the hierarchy of each organisation and horizontally across the different organisations involved in the project;
- responsibility and capability to perform assigned tasks at the expected time, within the expected budget, and with the expected quality;
- ability to restrain commercial prejudices and empathise with other participating organisations.

13.4.2.3 CI enabling values
The interviewees outlined the enabling values required to facilitate the CI efforts to include:

- trust;
- openness and transparency;
- humility;
- cooperation;
- respect for people;
- responsibility and integrity;
- empathy and responsiveness.

13.4.2.4 CI infusing values
In order to infuse the driving and enabling values into the culture and practices of multi-party projects, the following key sustaining values were outlined by the interviewees:

- commitment and leadership by the top management of all participating organisations in providing deliberate role modelling, teaching and coaching and the design of an environment for nurturing and promoting mutual learning;
- team-based organisational structure that foster better communication, improve respect and create trust;
- supporting infrastructure in the form of policy formulation and deployment, reward and recognition system, monitoring and measurement, learning and knowledge capture;
- shift in control system from inspection and policing to prevention and empowerment through systematic idea management;
- behavioural change programme with specific improvement targets and a clear reward system.

13.4.2.5 Barriers to achieving CI in LIFT projects
The key barriers identified by the interviewees, which if removed are capable of improving performance and facilitating CI in LIFT projects, include:

1. distrust/suspicion and lack of mutual understanding;
2. different *modus operandi;*
3. partners operating on different time-frames;
4. lack of clarity and communication of goals, roles and responsibilities;
5. lack of appropriate skills and competencies;
6. adversarial context.

13.5 The development of a generic continuous improvement framework for LIFT

Using the three-stage process-view of improvement developed by Atkin *et al.* (2003) and Barlow and Jashapara's (1998) argument that long-term relationships provide considerable opportunity for learning from project-to-project, thereby facilitating CI of products and services, a conceptual framework for achieving CI in long-term relationships is developed and shown in Figure 13.1.

The model determines what factors lead to the CI at each of the three stages of construction process in a sequential process flow from left to right, with the feedback and feed-forward loops distinguishing a one-off relationship from long-term relationships and indicating the commencement of other cycles. The model is generic in nature and can be applied in a variety of contexts. However, the achievement of desired results will depend on the identification of unique values and learning dynamics for different organisations and sectors. For example, the primary care sector is heavily influenced by rapidly changing demographies; clinical technologies and innovations; fashions; expectations; and increasing opportunities for different ways of working offered by advances in information and

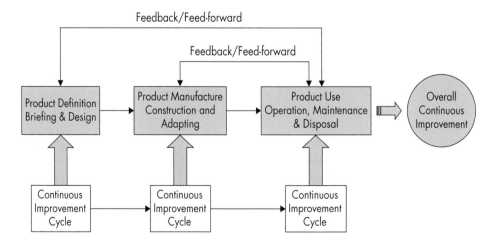

Figure 13.1   A conceptual model of continuous improvement in long-term relationships.

communication technologies and process redesign (Ibrahim and Price, 2005a, 2005b, 2006).

HaCIRIC (2007) described the health and social care system as one of the most complex and rapidly changing organisational and technical environments in any sector of the economy; involving multiple stakeholders that participate in care delivery, characterised with convoluted funding mechanisms and rapidly changing patterns of demand and use as well as government policies. These factors necessitate a shift to an approach that will facilitate dynamic continuous learning in order to accommodate the constantly evolving changes in a way that drive greater efficiency. This necessity is to ensure that the constructed primary care facilities are not only fit-for-purpose but are also fit-for-the-future.

The proposed framework is based on an underlying assumption that the achievement of gradual, consistent and ongoing CI requires focus on the organisation, the processes involved in the organisation, the employees working in the organisation, the external influences and the dynamic and complex interactions between the different actors within sector. Overall, the operational philosophy of the proposed framework is to assist organisations involved in long-term construction contracts in identifying performance gaps by analysing current operations, and through that, identifying the causes of gaps, and to generate and manage improvement activities targeted at closing the performance gaps from their previous and others' practices. In so doing, a full-blown model will include vertical, horizontal and longitudinal processes. The vertical processes include both the top-down strategy-driven process of goal-setting and deployment around commonly agreed improvement initiatives and the bottom-up process of reporting the results to sustain the initiatives through effective feedback mechanism. The horizontal processes involve dissemination and exchange of results and experience obtained from integrated cross-functional teams. The longitudinal processes involve alignment of the requirements, values and the working processes of the various stakeholders as well as feedback and feed-forward mechanisms of lessons learnt throughout the whole life cycle of the projects.

13.6 Application of CIf within LIFT procurement

The proposed framework is targeted to be implemented at the level of *LIFTCos*. Typical partners in each *LIFTCo*. include a private sector consortium of diverse specialties (selected from competitive tendering), local stakeholders (such as Primary Care Trusts and the Local Authorities) and Partnerships for Health (owned by the DoH). In operationalising the conceptual CI model (Figure 13.1), the three-dimensional perspective proposed by de Wit and Meyer (2005) for implementing strategic management is adopted. The three components are contextual analysis, CI strategy formation and CI implementation which are akin to de Wit and Meyer's (2005) strategic thinking, strategy formation and strategic change respectively. The three main components are discussed in the following section as a 10-step process. It should be noted that these components are not phases, stages or elements that can be understood in isolation, but interact in non-linear, intertwined and continuous fashion.

13.6.1 Contextual analysis

This involves the analysis of the current strategic position of the organisation to identify it strengths and weaknesses as well as the threats and opportunities in the external environment. The contextual factors are the key external underlying factors that affect the procurement of primary care facilities, such as clinical technologies and innovation, demographic changes, government policies, information and communication technologies etc. These factors interact within complex contexts involving multitude institutional and organisational stakeholders, sectors and components and determine the shape and pattern of the care system at both macro level (involving various care pathways and clinical procedures) and at the more micro level (reflected in the different life cycles of the various components of both the care system and the care environments). All these factors need to be reconciled with the other more sector-specific factors in order to achieve the desired results. The contextual analysis process would involve the following three steps:

1. External audit – involving assessment of each LIFT locality, wider health and construction industries, market analysis, competitor analysis, and the identification of key opportunities and threats offered and posed by the LIFT business environment.
2. Internal audit – involving the establishment of the strategic capability, strengths and weaknesses of the overall LIFT partnership through, for example, identification of core competencies and value chain analysis.
3. Some form of analysis of strengths, weaknesses, opportunities and threats (SWOT) to identify how the LIFT partnership's current capabilities are able to deal with the challenges of their specific local healthcare environments.

13.6.2 CI strategy formation

This involves the establishment of a strategic plan for institutionalising CI in the organisation. The institutionalisation of CI culture in organisations is being advocated through disciplined integration of learning into standard project management practice by:

*   enforcing project debriefings in all relevant LIFT project guidelines and policies, and integrating learning and knowledge goals at both micro and macro levels, that is into each LIFT project phase models and into the overall project goals and metrics of each LIFT scheme respectively. At the macro level, there

is a need to extend project outcomes beyond the physical and tangible results to incorporate the contribution of the project to the organisation's knowledge base;

- training and educating *LIFTCo.* project members about the importance of systematic debriefings and in undertaking a collective, interactive evaluation and analysis of experiences made by individual team members. The workshops should encourage the use of graphical platform, for example collecting and structuring the project experiences along a timeline (e.g. as a process map with mistakes, successes, insights etc.) and provision of the results in a poster format visible for all staff involved showing performance against targets;
- encouraging *LIFTCo.* project managers to lead by example and make project debriefings a strategic priority with clear roles and responsibilities.

CI strategy formation involves the following six steps:

1. Identification, engagement and analysis of the LIFT project stakeholders at various stages of the procurement in order to capture and understand their values and priorities. While the production of the Strategic Service Development Plans (SSDPs) can be said to facilitate the generation of robust requirements of the various stakeholders, their processing and management into project requirements and products suffer the same shortcomings as those of traditional client briefs. Consequently, more systematic stakeholder identification, engagement and analysis mechanisms are advocated to effectively capture the values and requirements of the diverse groups and systematically assess their claims, interests and power to influence decisions, such that jointly agreed strategies can be formulated to cover both planning and implementation of care services and facilities.
2. Development of a joint CI strategic plan by all LIFT stakeholders – setting missions, goals and objectives; clarifying policies and principles; and agreeing a shared CI strategy and alignment mechanism.
3. Development of jointly agreed processes required to achieve the harmonised strategy as well as the review points, success (performance) measures and training plan by all LIFT stakeholders.
4. Definition and assignment of roles, responsibilities and accountabilities for each process as well as the communication and reporting strategies related to all LIFT key stakeholders. The development of a procurement process map that shows clear information flows, deliverables, approval and review points, and identify the roles and responsibilities and the appropriate skills mix that are required to satisfy each of the process stages is advocated at both strategic and operational levels. Cross-functional diagrams (to show who is responsible for specific tasks and decisions, and the sequence of those actions) and Responsible, Accountable, Support, Consult and Inform (RASCI) charts (to show for each task who is responsible for carrying it out, who is authorised to approve work

or expenditures, who provides administrative or technical support, who should and can provide counsel and who should be kept informed) are commonly used in conjunction with jointly defined governing principles and operating guidelines. In addition, the process map should also indicate the members of each team or community of practice responsible for delivering each task, the deliverables, the task duration and a unique identification number for easy reference in other documentations and guidance notes.

The underlying principle for the allocation of roles and responsibilities is to ensure that each LIFT partner retains accountability for delivering the part of the project for which it has been selected (e.g. design, finance, construction, fabrication etc.) whilst at the same time signing up to the collective responsibility of successfully delivering the completed project. Therefore, the allocation of personnel, especially to key positions, should reflect individual corporate accountability and each of the key functional areas should be led by a person from the LIFT partner that is accountable for that function. The potential conflicts of interest that can arise when an individual has a dual role (such as member of SPB and PCT) should also be managed in the same light, that is accountability should be on the basis of original parent organisation. In order to reflect the collective responsibility of the LIFT partners, allocation of responsibilities should avoid duplication and on the basis of 'best for the job', while avoiding parochial protectionism. To correct the current situation, an 'as is' skills mix can be mapped so that the gaps can be identified and these should be filled preferably from partners or if necessary from outside the organisation or outsourced.

5. Formation of continuous improvement teams (CITs) and continuous improvement champions (CIC) group and the appointment of a continuous improvement facilitator (CIf) for each LIFT scheme. The appointment of a CIf to act as an independent and neutral facilitator who manages the debriefing processes under workshop settings (i.e. the workshop preparation, the workshop itself and its documentation) at the approved learning events is advocated. The CIf should be responsible for providing suitable tools and could be from within or outside the organisation, but preferably an outsider who understands the working processes and practices of all the individual parties involved in the project and has access to all the relevant project documents. Rather than traditional meetings, formal CITs could be formed on the basis of functional groups within each LIFT and the learning events can concur with the stage-gate meetings of each CIT. The CITs are responsible for generating ideas for improvement termed 'potential continuous improvement ideas' (PCIIs).

6. Designing an appropriate organisational structure, management systems, operational policies and procedures, action plans, short-term budgets/resources allocation and information systems for implementing above mentioned steps 3–5 within each LIFT scheme.

13.6.3 CI implementation

This is concerned with the translation of CI strategic plan into action, and comprises the essential practices required at various procurement stages to facilitate CI, collectively indicating 'what' and 'how' the LIFT partner organisations need to do to achieve CI. These comprise of behavioural aspects, internal processes as well as tools and techniques. The implementation process involves the following step:

1. generation of PCIIs by the CITs at each of the review points identified in step 3 of CI strategy formation. During each learning event, a four-step PDCA cycle is proposed to be used to facilitate the process.

Neve (2003) constructed a toolkit that serves as a dynamic mechanism for forcing individuals to continually rethink their actions, their implicit assumptions, their relations to others and to their environment. Such a toolkit can be adapted for use in during learning events in the procurement of primary care facilities. Periodically, a CIC team comprising of senior employees from cross-organisational disciplines in each LIFT scheme and with responsibility of evaluating all the PCIIs should meet along with the CIf to agree on the ideas that can add value towards the achievement of the organisation's strategic plan and also to validate the lessons and knowledge captured by the CITs along with their associated contexts, before storage and dissemination.

13.7 Conclusions

The Department of Health, which is responsible for maintaining the overall health of people living in England through the National Health Service (NHS), introduced the LIFT initiative to reverse the declining state of primary care infrastructure. Although the initiative can be said to be moving towards delivering the expected targets, the attainment of the contractual requirements for the demand and supply sides to continuously improve performance under the LIFT scheme had been severally shown to remain intangible. Accordingly, a research was launched to: explore the CI concept in general; identify the requirements for implementing CI in LIFT; develop a generic framework for achieving CI; and identify the key application challenges with the LIFT procurement initiative.

The methodology adopted for the research involved review of relevant literature and semi-structured interviews with senior managers of organisations involved in three LIFT schemes. Literature syntheses indicated that the principal prerequisites for CI success include: a clear strategic framework, which is clearly communicated to all employees together with related long-term and short-term targets and milestones, and the exploration of *driving* and *enabling* values. The interviews revealed that CI implementation and applications remains diverse and subject to varied understandings and interpretations from both the demand and supply sides. Nonetheless, a common theme that can be implied from this research is that CI concept is based on effort to improve the performance of every facet of an organisation, including processes, products and services. However, a common understanding and explicit set of tools and techniques are needed for any meaningful progress in enshrining it in the culture and practices of any industry. The importance of values, behaviours and culture to the success of CI has been widely recognised, and this research investigated the values that motivate and sustain CI effort. The interviews also identified the success factors as well as the driving, enabling and infusing values for achieving CI under LIFT initiative.

The proposed framework comprises the following three main components: contextual analysis (which involves the analysis of the current strategic position of the organisation such as clinical technologies and innovation, demographic changes, government policies, information and communication technologies); CI strategy formation (which involves the formulation of a strategic plan for institutionalising CI in the organisation, including operational guidelines at individual, team and organisational levels); and CI implementation (which is concerned with the translation of the CI strategic plan into action and include the essential practices required at various procurement stages to facilitate CI, collectively indicating 'what' and 'how' organisations need to do to achieve CI and comprise of behavioural aspects, internal processes as well as tools and techniques). Although no empirical validation of this framework has been carried out, it presents a theoretical starting point that is hoped would ensure that the constructed LIFT facilities are not only fit-for-purpose but are also fit-for-the-future.

References

Ahmad, I.U. and Sein, M.K. (1997) Construction project teams for TQM: a factor-element impact model, *Construction Management and Economics*, **15**, 457–467.

Anumba, C.J., Baugh, C. and Khalfan, M.A.M. (2000) Organisational structures to support concurrent engineering in construction, *Industrial Management Data System*, **102**, 260–270.

Atkin, B., Borgbrant, J. and Josephson, P. (2003) *Construction Process Improvement*, Blackwell Science, Oxford, UK.

Baggot, R. (2004) *Health and Health Care in Britain*, 3rd edn, Palgrave Macmillan, Basingstoke.

Barlow, J. and Jashapara, A. (1998) Organisational learning and inter-firm 'partnering' in the UK construction industry, *The Learning Organisation*, **5**(2), 86–98.

Berger, A. (1997) Continuous improvement and Kaizen: standardisation and organisational design, *Integrated Manufacturing Systems*, **8**(22), 110–117.

Bessant, J. and Caffyn, S. (1997) High involvement innovation through continuous improvement, *International Journal of Technology Management*, **14**(1), 7–28.

Bessant, J., Caffyn, S., Gilbert, J., Harding, R. and Webb, S. (1994) Rediscovering continuous improvement, *Technovation*, **14**(1), 17–29.

Bresnen, M.J. (1990) *Organising Construction: Project Organisation and Matrix Management*, Routledge, London.

Bresnen, M., Edelman, L., Newell, S., Scarborough, H. and Swan, J. (2003) Social practices and the management of knowledge in project environments, *International Journal of Project Management*, 21(2), 157–166.

Caffyn, S. (1999) Development of a continuous improvement self-assessment tools, *International Journal of Operations and Production Management*, 19(11), 1138–1153.

Chinowsky, P., Molenaar, K. and Realph, A. (2007) Learning organisations in construction, *Journal of Management in Engineering*, 23(1), 27–34.

Cooper, R., Aouad, G., Lee, A., Wu, S., Fleming, A. and Kagioglou, M. (2005) *Process Management in Design and Construction*, Blackwell Publishing, London.

Davey, C.L., Powell, J.A., Cooper, I. and Powell, J.E. (2004) Innovation, construction SMEs and action learning, *Engineering, Construction and Architectural Management*, 11(4), 230–237.

De Wit, B. and Meyer, R. (2005) *Strategy Synthesis: Resolving Strategy Paradoxes to Create Competitive Advantage*, 2nd edn, Thomson Learning, London.

DoH. (2000) *The NHS Plan – A Plan for Investment, A Plan for Reform*, Department of Health, London.

DoH (2001) *Public Private Partnerships in the NHS: Modernising Primary Care in the NHS-Local Improvement Finance Trust (NHS LIFT) – Prospectus*, Department of Health and Partnerships, London, UK.

Egan, J. (1998) *Rethinking Construction*, Department of Trade and Industry, London.

Errasti, A., Beach, R., Oyarbide, A. and Santos, J. (2007) A process for developing partnerships with subcontractors in the construction industry: an empirical study, *International Journal of Project Management*, 25(3), 250–256.

Ferng, J. and Price, A.D.F. (2005) An exploration of the synergies between Six Sigma, Total Quality Management, Lean Construction and Sustainable Construction, *International Journal of Six Sigma and Competitive Advantage*, 1(2), 167–187.

Fernie, S., Green, S.D., Weller, S. and Newcombe, R. (2003) Knowledge sharing: context, confusion and controversy, *International Journal of Project Management*, 21(3), 177–187, ISSN 0263-7863.

Fernie, S., Weller, S., Green, S.D., Newcombe, R. and Williams, M. (2001) Learning across business sectors: context, embeddedness and conceptual chasms, *Proceedings of 17th Annual ARCOM Conference*, University of Salford, UK, pp. 557–565, ISBN 0 9534161 6X.

Fernie, S., Weller, S., Green, S.D., Newcombe, R. and Williams, M. (2002) Knowledge sharing: a softly-softly approach. In T.M. Lewis (ed.), *Proceedings of CIB W-92 Symposium: Procurement Systems and Technology Transfer*, University of the West Indies, Trinidad & Tobago, pp. 555–576.

Gallagher, M., Austin, S. and Caffyn, S. (1997) *Continuous Improvement in Action: The Journey of Eight Companies*, Kogan Page, London.

Garfield, C. (1986) *Peak Performers: The New Heroes of American Business*, William Morrow and Company, New York.

Green, S., Newcombe, R., Fernie, S. and Weller, S. (2004) *Learning Across Business Sectors: Knowledge Sharing between Aerospace and Construction*, BAE Systems, UK.

Green, S.D., Newcombe, R., Williams, M., Fernie, S. and Weller, S. (2002) Supply chain management: a contextual analysis of aerospace and construction. In T.M. Lewis (ed.), *Proceedings of CIB W-92 Symposium: Procurement Systems and Technology Transfer*, University of the West Indies, Trinidad & Tobago, pp. 245–261.

Grimsey, D. and Graham, R. (1997) PFI in the NHS, *Engineering, Construction and Architectural Management*, 4(3), 215–231.

Health and Care Infrastructure Research and Innovation Centre (HaCIRIC). (2007) About HaCIRIC, available at: http://www.haciric.org/about-haciric [accessed 12 May 2007].

Hill, F.M. (1996) Organisational learning for total quality management through quality cycles, *TQM Magazine*, 8(6), 53–57.

Holmes, J., Capper, G. and Hudson, G. (2006) Public private partnerships in the provision of health care premises in the UK, *International Journal of Project Management*, 24(7), 566–572.

Hudson, G., Capper, G and Holmes, J. (2003) The implication of PFI on health care premises engineering design, durability, and maintenance, *The Institution of Mechanical Engineers Conference Transactions*, pp. 33–40.

Hyland, P., Mellor, R., Sloan, T. and O'Mara, E. (2000) Learning strategies and CI: lessons from several small and medium Australian manufacturers. *Integrated Manufacturing Systems*, 11(6), 428–436.

Ibrahim, A.D. and Price, A.D.F. (2005a) Conceptualising a continuous improvement framework for long-term contracts: a case study of NHS LIFT. In: C. Egbu and M. Tong (eds), *Proceedings of 2nd Scottish Conference for Postgraduate Researchers of the Built and Natural Environment (PRoBE)*, Glasgow Caledonian University, UK, 16th–17th November 2005, pp. 229–241, ISBN 1-903661-82-X.

Ibrahim, A.D. and Price, A.D.F. (2005b) Impact of social and environmental factors in the procurement of healthcare infrastructure. In: C. Egbu and M. Tong (eds), *Proceedings of 2nd Scottish Conference for Postgraduate Researchers of the Built and Natural Environment (PRoBE)*, Glasgow Caledonian University, UK, 16–17 November 2005, pp. 217–228, ISBN 1 903661 82 X.

Ibrahim, A.D. and Price, A.D.F. (2006) Public private partnerships and sustainable primary healthcare facilities in Nigeria. In: I.A. Okewole, A. Daramola, A. Ajayi, K. Odusami and O. Ogunba (eds), *Proceedings of International Conference on the Built Environment*, Covenant University, Nigeria, January 24–26, pp. 221–227.

Ibrahim, A.D., Price, A.D.F. and Dainty, A.R.J. (2008) Is the Local Improvement Finance Trust (LIFT) procurement initiative delivering the expected economies of scale? Results from three case studies. In: A.R.J. Dainty (ed.), *Proceedings of ARCOM 2008 Conference*, Cardiff, UK, September 3–4, 2008.

Imai, K. (1986) *Kaizen*, Random House, New York.

Jabnoun, N. (2001) Values underlying continuous improvement. *The TQM Magazine*, 13(6), 381–387.

Juergensen, T. (2000) *Continuous Improvement: Mindsets, Capability, Process, Tools and Results*, The Juergensen Consulting Group Inc., Indianapolis, IN.

Kerrin, M. (1999) Continuous improvement capability: an assessment within one case study organisation, *International Journal of Operations and Production Management*, 19(11), 1154–1167.

Kumaraswamy, M.M. (1998) Reconstructing the team, *Proceedings of the CIB W89 International Conference on Building Education and Research (BEAR'98)*, CIB, Brisbane, Australia, pp. 262–272.

Latham, M. (1994) *Constructing the Team*, Department of the Environment, HMSO, UK.

Lillrank, M. (1995) The transfer of management innovations from Japan, *Organisation Studies*, 16(6), 971–989.

Maqsood, T., Walker, D.H.T. and Finegan, A.D. (2007) Facilitating knowledge pull to deliver innovation through knowledge management: a case study, *Engineering, Construction and Architectural Management*, 14(1), 94–109.

National Audit Office (NAO). (2005) *Department of Health: Innovation in the NHS – Local Improvement Finance Trusts*, Report by the Comptroller and Auditor General HC 28 Session 2005–2006, National Audit Office, London.

Neve, T.O. (2003) Right questions to capture knowledge, *Electronic Journal of Knowledge Management*, 1(1), 47–54.

Ngowi, A.B. (2000) Construction procurement based on current engineering principles, *Logistic Information Management*, 13(6), 361–368.

Nwakoby, B.A. (2004) *PHC Services and Quality of Care*, National Stakeholders Consultative Meeting on Primary Health Care in Nigeria organised by the National Primary Health Care Development Agency (NPHCDA) at Women Development Centre, Abuja, July 22–23.

Oakland, J.S. (1995) *Total Quality Management*, 2nd edn, Butterworth-Heinemann, Oxford.

Oakland, J. (1999) *Total Organisational Excellence – Achieving World-Class Performance*, Butterworth-Heinemann, Oxford.

Pasquire, C.L. and Connolly, G.E. (2002) Leaner construction through off-site manufacturing, *Proceedings of the 10th Annual Conference of the International Group for Lean Construction*, Gramado, Brazil, August 2002, pp. 263–276.

Pettigrew, A. and Whipp R. (1991) *Managing Change for Competitive Success*, Basil Blackwell, Oxford.

Powell, J.A. (1999) Action learning for continuous improvement and enhanced innovation in construction, *Proceedings of 7th IGLC Conference*, University of California, Berkeley, CA, July 26–28.

Robinson, A. (1991) *Continuous Improvement in Operations*, Productivity Press, Cambridge, MA.

Savolainen, T (1998) Managerial commitment process in organizational change: findings from a case study, *Academy of Strategic and Organizational Leadership Journal*, 2(2), 1–12.

Schindler, M. and Eppler, M.J. (2003) Harvesting project knowledge: a review of learning methods and success factors, *International Journal of Project Management*, 21, 219–228.

Slaughter, E.S. (1998) Models of construction innovation, *Journal of Construction Engineering and Management*, 124(2), 226–231.

Slaughter, E.S. (2000) Implementation of construction innovations, *Building Research and Information*, 28(1), 2–17.

Steel, J. and Murray, M. (2004) Creating, supporting and sustaining a culture of innovation, *Engineering, Construction and Architectural Management*, 11(5), 316–322.

Suzaki, K. (1987) *The New Manufacturing Challenge*, The Free Press, New York.

Tague, N.R. (2004) *The Quality Toolbox*, 2nd edn, ASQ Quality Press, Milwaukee, WI.

Tan, H.C., Carrillo, P.M., Anumba, C.J., Bouchlaghem, N.M., Kamara, J.M., and Udeaja, C.E. (2007) Development of a methodology for live capture and reuse of knowledge in construction, *Journal of Management in Engineering*, 23(1), 18–26.

Tennant, C., Warwood, S.J. and Chiang, M.M.P. (2002) A continuous improvement process at Seven Trent Water, *The TQM Magazine*, 14(5), 284–292.

Thomas, G. and Thomas, M. (2005) *Construction Partnering and Integrated Teamworking*, Blackwell Publishing Ltd., Oxford.

Towill, D.R. (2003), Construction and the time compression paradigm, *Construction Management and Economics*, 21, 581–591.

Whetherill, M., Rezgui, Y., Lima, C. and Zarli, A. (2002) Knowledge management for the construction industry: the e-COGNOS Project, *ITcon*, 7, Special Issue: ICT for Knowledge Management in Construction, 183–196.

Williams, T.M., Ackermann, F.R., Eden, C.L. and Howick, S. (2001) The use of project post-mortem, *Proceedings of the Project Management Institute (PMI) Annual Symposium 2001*, Nashville, TN, November.

Performance Management in the Context of Healthcare Infrastructure

14

Therese Lawlor-Wright and Mike Kagioglou

This chapter looks at the concept of performance management and how it has been interpreted in organisations responsible for delivering healthcare infrastructure. The performance-based building concept assesses the quality of a building in terms of how well it meets user requirements. Increasingly, public–private partnerships, for example private finance initiative (PFI), Local Improvement Finance Trust (LIFT) transfer responsibility to private sector partners to maintain the performance of the infrastructure and to supply support services such as facilities management (FM).

Increasingly, NHS organisations are competing in a healthcare marketplace. Healthcare choices are often made on the basis of previous patient experiences and public information on the performance of NHS trusts. Many surveys have shown that infrastructure and FM services (cleaning, security, food) have a very important influence on the patient experience. Therefore, infrastructure providers have a significant impact on the ability of the healthcare organisation to compete with other providers.

The main conclusion from this work is that there is a need for a holistic concept of performance-based specification (PBS) in order to ensure that infrastructure makes maximum contribution to the performance of the healthcare organisation. Such a definition would emphasise the importance of FM and foster closer co-operation between organisations involved in healthcare infrastructure.

14.1 Introduction

In an increasingly competitive global marketplace, organisations have sought to measure and improve their performance to achieve competitive advantage and survive. The concept of performance management refers to an active process of measuring and taking action in order to improve performance. According to the Oxford English Dictionary, 'Performance' refers to the doing of an action and the quality of how the action is performed. The competence or effectiveness of the person or organisation in carrying out the action is often inferred. Performance, in financial terms, may be used to refer to the profitability of an investment. However, profit is not the only measure of how well an organisation is performing. Most financial reports are 'lagging' indicators in that financial irregularities show up after a disturbance in performance, when it may be too late to take corrective action. Thus, there is a need for a set of measures which are 'leading' and warn the company of unusual activity, so that action can be taken before there is financial impact. Much of the research in the area of organisational performance management focuses on the use of measures to achieve business strategic objectives (Neely *et al.*, 2005).

In the field of architecture and design, the concept of building performance has been used to describe whether a building fulfils its design requirements. Preisser and Vischer (2004) define building performance evaluation as 'the process of systematically comparing the actual performance of buildings, places and systems to explicitly documented criteria for their expected performance'.

A healthcare system consists of a healthcare organisation and healthcare facilities (buildings, equipment, environment). The purpose of research being undertaken at the Health and Care Infrastructure Research Centre (HaCIRIC) at the University of Salford is to investigate the components of healthcare performance both in terms of services and infrastructure and how they are related. This chapter describes the initial stages of the research and the literature sources that inform it.

The chapter is structured as follows. Section 14.2 looks at the field of organisational performance management and the most commonly used methods and tools. Section 14.3 looks particularly at how performance has been interpreted in the design and refurbishment of infrastructure. Section 14.4 looks at performance of healthcare facilities and how this has been interpreted, particularly in the context of the UK National Health Service (NHS). Section 14.5 discusses the relationship between facilities performance and healthcare organisational performance and proposes further research to date in this area.

14.2 Organisational performance measurement systems

A performance measurement system (PMS) is the set of measures used to quantify both the efficiency and effectiveness of actions (Neely, 2005). Establishing a set of measures indicates to the organisation what the priorities are, since it focuses attention on improving the measures. For this reason, many authors have emphasised the importance of having measures that are congruent with the organisational strategy. For example, Rangone (1997) states that any PMS needs to have strategy as a main input, so that any results coming out of the system can be used to evaluate the extent to which the organisation has met its strategic goals.

Recognising that financial measures fail to capture all aspects of business performance, the Balanced Scorecard is an alternative model proposed by Kaplan and Norton (1991). In addition to the financial perspective, managers are asked to consider how their organisation is performing from the perspective of the customer and how the processes within the organisation are working. In addition, because the company needs to sustain its performance in the long term, managers are asked to consider how the company is performing with respect to staff learning and development.

In 1997, The Harvard Business Review listed the Balanced Scorecard as one of the 75 most influential ideas of the 20th century (Bible *et al.*, 2006). The first step in its implementation is to create a strategy map, showing the management's view of success and how it is achieved. The process of creating a scorecard and translating it into measures for departments and teams involves everybody in achieving the organisational strategy. Although performance measurement techniques were developed for use in industry, they have been adapted for use in the public sector. The focus of such measures is often to identify if society is achieving 'value for money' from the public body and to use measures to improve the efficiency and effectiveness of service delivery (De Bruijn, 2002).

The results of performance measurement tell what happened, but not why it happened or what to do about it. Thus, the role of the performance management process is to extract insight from performance data to support decision-making and bring about performance improvements (Bourne *et al.*, 2003; Neely and Al Najjar, 2006).

14.3 Building performance assessment

The required functionality of a building is a primary focus of the design team, in this context – 'performance' is used to denote how well the building meets user requirements (Preisser and Vischer, 2004). Becker (1999) describes the 'performance' approach as working in terms of end requirements rather than their means of delivery. The concept of 'performance-based building' is based on:

1. translating human needs into user requirements;
2. translating user requirements into technical requirements and quantitative criteria;
3. using these requirements in the life cycle of the building.

Thus, the performance of a building or its subsystems is assessed as how well it meets the user requirements.

Three levels of building performance have been identified by Preisser and Vischer (2004), roughly paralleling the hierarchy of human needs proposed by Maslow (1948). Maslow proposed that individuals have a hierarchy of needs ranging from the basic (food, shelter, security) through emotional and social needs and reaching a pinnacle at self-actualisation. His insight was that individuals need to have their lower level needs met before they manifest higher-level needs.

A wide range of methods and frameworks for performance measurement of buildings have been proposed (Mc Dougall *et al.*, 2002). These range from detailed technical assessments to surveys of occupied space and user satisfaction (Then, 2004). Post-occupancy evaluation (POE) is the evaluation of buildings in a systematic manner after they have been built and occupied for some time (Preisser and Vischer, 2004). It measures the extent to which the end-user requirements and goals have been met by the building. Measures relate to organisational performance, worker satisfaction and productivity levels as well as measures of building performance such as lighting levels and adequacy of space.

During the life of the building, FM is responsible for maintaining the building performance and adapting it to meet new user requirements. The functions of facilities managers include: property strategy, space management, communications infrastructure and building maintenance (BIFM, 2007). FM often assumes responsibility for other functions including security, catering and cleaning. These correspond with lower levels on the Maslow hierarchy of needs, satisfying the physiological needs of patients, visitors and staff. Barrett (1995) notes that the role of FM is to 'create an environment that strongly supports the objectives of the organisation'. This emphasises that the performance of any organisation is dependant on the performance of the FM organisation that supports it.

14.3.1 Performance of healthcare facilities

Private finance initiative (PFI) is now used as the main vehicle for delivering additional resources to the healthcare sector (Akintoye and Chinyio, 2005) PFI enables a private partner to build a facility to the output specifications agreed with the public agency, operate the facility for a specified time period and then transfer the facility to the public sector client. The nature of health PFI facilities includes new build, conversions, redevelopment and modernisation schemes (NHS Executive, 1999). LIFT schemes are used to build or refurbish primary care premises and lease back on favourable terms to primary health service providers (DoH, 2005). Akintoye and Chinyio (2005) note that in many PFI schemes including LIFT, the private sector partner is being used to provide support services such as building maintenance, cleaning, administration and security. Since PFI transfers risks to the private sector partner, this has led to this partner being responsible for FM to ensure the continuous availability of the facility.

There are three sub-activities involved in delivering performance infrastructure:

1. delivery of new or refurbished infrastructure to meet user requirements;
2. maintenance of existing infrastructure in an appropriate condition to support the users;
3. delivery of services which are associated with infrastructure to support the users (e.g. cleaning, security).

This chapter now looks at how healthcare infrastructure performance is considered during each of these stages.

14.3.2 Assessing performance at the design stage

Various tools have been developed to allow NHS Trusts to evaluate their existing buildings and to assess new designs. The two main tools in this respect are AEDET and ASPECT. The AEDET Evolution toolkit (DoH, 2007c) has been developed to help NHS Trusts to determine and manage design requirements for new buildings and to assess the strengths and weaknesses of existing buildings. It is part of the guidance for procurement schemes for NHS buildings (including ProCure21, PFI and LIFT). The tool has three main sections addressing building impact, build quality and functionality. Each section consists of 10 qualitative assessment criteria and the tool is normally completed by achieving consensus on the ratings amongst a group of stakeholders. The contents of the sections are described in the following list:

1. *Impact:* The extent to which the building creates a sense of place and contributes positively to the lives of those who use it and are its neighbours.
2. *Build Quality: Performance* – This deals with the physical components of the building – is it soundly built, reliable and easy to operate? Will it last well and is it sustainable? Has the potential disruption in the construction process been minimised.
3. *Functionality:* Deals with how well the building serves its primary purposes and the extent to which it facilitates or inhibits the activities of people who carry out the functions inside and around the building.

Using these tools and frameworks allows an organisation to assess the performance of existing building and prioritise areas for improvement in new or refurbished healthcare facilities.

The ASPECT toolkit (NHS Estates, 2008) scores new and existing designs from the point of view of staff and patient experience. The assessment criteria used related to eight topics: privacy and dignity; views; access to nature and outdoors; comfort (lighting, sound, temperature); legibility (wayfinding); facilities (bathrooms, entertainment venues, worship, relaxation, visitors, availability of snacks and drinks) and appearance.

Design teams use these tools to assess if their requirements have been met. However, the range of users of health buildings is wide and their needs change over time. There are also many additional stakeholders in the community, especially when (as in LIFT schemes) the health building is part of the regeneration of a particular area.

14.3.3 Assessing performance at operational stage

Once a healthcare facility becomes operational, the FM function is responsible for ongoing performance management and building maintenance and increasingly for providing support services.

Several authors have proposed metrics for assessing the performance of healthcare buildings and the FM function. Friedriksen (2002) proposes a system of key performance indicators (KPIs) for hospital FM and incorporates these in a software tool. Pullen *et al.* (2000) proposes seven KPIs that provide benchmarks for asset management of medical facilities. Lavy and Shohet (2004) have proposed a system using four KPIs for hospital maintenance departments. These indicators are further described in Chapter 16. The indicators may be used to assess clinics and primary care facilities and to compare their performance with other clinics or over a given time period.

The UK Department of Health identified seven main target areas for performance monitoring of NHS Estates: safety, cleanliness, food service, linen services, maintenance, environment and communications (DoH, 2007a). A perception survey has been developed to score the quality of access, arrival, parking, reception, waiting areas, moving around, convenience, comfort, appearance, cleanliness, noise, security and treatment facilities before and after modernisation. In addition, targets have been set and budget allocated to eradicate the backlog in health and safety activities (DoH, 2000).

Currently, each NHS trust completes an estates return and information collection (ERIC) each year with performance measures for their estates (DoH, 2007b). The type of data included in the ERIC return is shown in Table 14.1 (DoH, 2007b; HEFMA, 1999). Measures include key areas such as domestic services, catering, maintenance, safety and security, laundry, the environment and telephones and communication (HEFMA, 2001).

In healthcare facilities, cleanliness is crucial and the key to reducing healthcare acquired infections. A detailed performance assessment framework has been designed to assess hospital cleanliness levels (NHS, 2006). This is intended to be used by an external assessment team and uses a five-point scoring scale. The teams, known as Patient Environment Action Teams (PEAT) are also used to assess food quality in NHS Trusts (NHS, 2007). Because of the disestablishment of NHS Estates, these functions were taken over by the NHS National Patient Safety Agency.

Table 14.1   National performance framework for NHS estates HEFMA (1999).

Indicators (all per square metre)

Space efficiency	Income
Relates the assets base and occupancy costs to output	Activity
	Asset value
	Occupational costs
Productivity	Asset value
Measures cost of owning asset	Capital charges
	Total backlog
	Rent and Rates
Cost occupancy	Afford ratio = income/occupancy cost
Provides a benchmark for occupancy cost	Occupancy cost = rent and rates /m^2
	Energy and utility cost
	Maintenance costs
	Capital charges
Estate quality	Asset value
Provides a balanced view of overall condition of assets related to value and age	Depreciation
	Health and safety and fire backlog
	Physical backlog
Asset deployment	Land value
Measures how land and equipment is deployed	Equipment value
	Building value
	Capital charges

14.4 Contribution of infrastructure to performance of healthcare organisation

In applying performance management concepts to healthcare organisations, the main focus has been on measuring the performance of hospitals. Kocakulah and Austill (2007) note that despite its widespread use in industry, the healthcare sector has been slow to adopt the Balanced Scorecard approach. These authors point out that many hospitals do not see the need to implement Balanced Scorecards since they already have to track many non-financial performance measures required by accreditation and government agencies.

A World Health Organisation European group has been formed to build and validate a flexible and comprehensive hospital performance assessment model and to allow participants to benchmark their performance against that of other hospitals (Shaw, 2003; WHO, 2003a, 2003b). WHO (2003a) reports on different measures of performance in hospitals around the world and reports on consensus from a range of experts on the key elements of hospital performance as shown in Table 14.2.

Considering Table 14.2, infrastructure contributes in many ways to all of the performance criteria identified.

- Clinical effectiveness often relies on having specialised diagnostic and treatment equipment and having a clinically clean environment. In addition, patients need safe and secure spaces in which to recover after treatment.
- Patient Centredness – Patient satisfaction and the quality of the patient experience as well as their access to amenities is assessed using design performance assessment tools and measured in patient satisfaction surveys.
- Production Efficiency – This is a focus of measures of the health service itself and of the systems used to measure record infrastructure condition. Healthcare infrastructure is a major investment of resources for a healthcare organisation, making more efficient use of these resources is an important part of improved healthcare performance.
- Safety – Ensuring the safety of staff and patients is often a function of FM organisations supporting the health service.
- Staff – Expenditure on non-clinical staff is a very significant part of the budget for the health service. In an analysis of costs in a US hospital, it was found that 31.5% of the total running costs were for support services such as

Table 14.2 Key dimensions of hospital performance (WHO, 2003a).

Dimension	Including
Clinical effectiveness	Technical quality, evidence based practice and organisation, health gain, outcome (individual and population)
Patient centredness	Responsiveness to patients, client orientation, (prompt attention, access to social support, quality basic amenities, choice of provider), patient satisfaction, patient experience (dignity, confidentiality, autonomy, communication)
Production efficiency	Resources, financial (financial systems, continuity, wasted resource), staffing ratios, technology
Safety	Patients and providers, structure, process
Staff	Health, welfare, satisfaction, development (e.g. turnover, vacancy, absence)
Responsive governance	Community orientation (answer to needs and demands), access, continuity, health promotion, equity, adaptation abilities to the evolution of the population's demands (strategy, fit)

utilities, employee benefits, building maintenance and housekeeping salaries (Roberts *et al.*, 1999).

• Responsive Governance/Community Focus – Increasingly healthcare infrastructure has other roles within the community it serves providing a venue for leisure facilities and a focus for regeneration in deprived areas (e.g. LIFT schemes). Locating services close to the community it serves can increase health awareness and improve access as well as reducing cost. For example in the 'Closer to Home' initiative, the Cumbria PCT identifies that managing and reducing the demand for hospital services can give a saving of £20m over three years (Cumbria NHS PCT, 2007).

In assessing the performance of infrastructure, two aspects must be considered. First there is the performance of the infrastructure itself, next there is the contribution of infrastructure to the health service performance. This contribution seems to have little recognition in health service management. For example, Gelnay (2002) and Payne and Rees (1999) note that FM is under-represented in the top-level hospital management. Thus, issues related to FM may not be considered when making key design decisions that have major implications for hospital costs. Despite the evidence that facilities are important to overall performance of healthcare organisations, the links are not clearly described and there is no apparent connection between PMSs used to assess the organisation and its facilities.

Infrastructure performance is essential to the smooth running of the health service and in satisfying the needs of healthcare staff, patients and visitors for a safe and secure environment. In a patient survey reported in the NHS Plan (DoH, 2000), 3 of the top 10 priorities for the health service were facilities issues: cleanliness, hospital food and a safe, warm and comfortable environment (Cole, 2004). Baldwin (2005) reports that patient use subjective assessments of the environment (ease of parking, facilities for visitors and perceived cleanliness) to make their healthcare choices. The impact of the hospital environment on patient choice is also reported by Coulter *et al.* (2004) and Miller and May (2006). Andaleeb (1998) reports that facility quality is a key factor influencing overall patient satisfaction. Hong Kong Authorities have also concluded that hospital domestic support services have a major impact on patient comfort and quality of care (Kam-Shim, 1999).

Figure 14.1 illustrates a holistic concept of healthcare organisational performance indicating how the performance of infrastructure and FM service is a major component of overall performance and underpins clinical performance and how health services are delivered. Infrastructure performance is highly dependent on design and build quality, these issues are addressed by approaches such as 'performance-based design'.

Then (2004) suggests the problem is not in finding building performance indicators but in selecting the right performance indicators from the myriad of possible

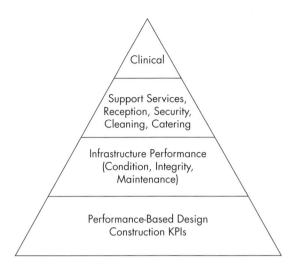

Figure 14.1   The components of hospital performance.

ones. As recommended by many authors in the field of performance management, measures for different parts of an organisation (or their external partners) should be related to the overall strategy that the organisation is trying to achieve. By aligning measures and objectives with those of the healthcare organisation, improved building and FM performance can contribute to the overall performance of the health service.

In PFI, miscommunication between public and private sector partners leads to under-performance. For example, Liyanage and Egbu (2005, 2006) point out that lack of communication between FM and clinical staff risks the cleanliness of hospitals. There is a need to capture best practice on healthcare FM and to share knowledge both within the NHS and with private sector partners providing FM services (HEFMA, 2003).

14.5 Conclusions

Performance measurement and management is assuming ever increasing importance in healthcare in the UK. In PFI contracts, risk is transferred to private sector partners making them responsible for performance on construction projects and in FM. Effective healthcare delivery now depends largely on effective collaboration between public and private sector partners. There is a need for integrated measures to be developed and for these to be used to raise the profile of the FM function within the healthcare organisation.

In healthcare, the focus is on measuring health outcomes. In FM, it is on building performance and the costs of maintaining the building and estates. There is a need to arrive at a common understanding of what constitutes performance and how performance measures at facilities level are related to the overall goals of the health-care organisation. This would facilitate co-operation and a common understanding between different parts of the organisation.

Attention gets devoted to what is measured in order to improve. However, some areas are more difficult to measure, for example measurement of health outcomes is very challenging. Some measures may encourage activity and organisational busy-ness rather than effectiveness and this must be taken into account when setting them. In order to 'work smarter' intra-organisational knowledge sharing must be taken into consideration. There is a need to further develop means of capturing the constituents of good and poor performance in order to facilitate learning and spread best practice.

14.6 Acknowledgements

This work was funded by the Engineering and Physical Sciences Research Council and conducted through the Salford Centre for Research and Innovation and the Health and Care Infrastructure Research and Innovation Centre.

References

Akintoye, A. and Chinyio, E. (2005) 'Private Finance Initiative in the Healthcare Sector: Trends and Risk Assessment', *Engineering, Construction and Architectural Management*, Vol. 12, No. 6, pp. 601–616.

Andaleeb, S. S. (1998) 'Determinants of Customer Satisfaction with Hospitals: A Managerial Approach', *International Journal of Health Care Quality Assurance*, Vol. 11, No. 6, pp. 217–223.

Baldwin, E. (2005) 'Patient Choice: Pick and Mix', *Health Service Journal*, Vol. 115, No. 5940, p. 38.

Barrett, P. (1995) '*Facilities Management: Towards Best Practice*', Blackwell Science, Oxford.

Becker, R. (1999) 'Research and Development Needs for Better Implementation of the Performance Concept in Building', *Automation in Construction*, Vol. 8, No. 4, pp. 525–532.

Bible, L., Kerr, S., and Zanini, M. (2006) 'The Balanced Scorecard: Here and Back', *Management Accounting Quarterly*, Vol. 7, No. 4, pp. 18–23.

BIFM. (2007) 'British Institute of Facilities Management', available at http://www.bifm.org.uk/bifm/about/facilities, accessed 22nd November 2007.

Bourne, M., Franco, M., and Wilkes, J. (2003) 'Corporate Performance Management', *Measuring Business Excellence*, Vol. 7, pp. 15–21.

Cole, A. (2004) 'Better Late than Never', *FM World*, pp. 18–19.

Coulter, A., Henderson, L., and Le Maistre, N. (2004) '*Patients Experience of Choosing Where to Undergo Surgical Treatment*', Picker Institute Europe, Oxford.

Cumbria NHS PCT. (2007) 'Closer to Home An NHS Consultation on Providing More Healthcare in the Community in North Cumbria', available at http://closertohome.org.uk/docs/PCTWeb.pdf, accessed 22nd May 2008.

De Bruijn, H. (2002) '*Managing Performance in the Public Sector*', Routledge, Taylor and Francis, London, ISBN 0-415-30037-1.

DoH, Department of Health. (2000) 'The NHS Plan: A Plan for Investment, A Plan for Reform', The Stationery Office, London, Chapter 6, p. 59.

DoH, Department of Health. (2005) 'Sustainable Development: Environmental Strategy for the National Health Service', The Stationery Office, London, available at http://www.dh.gov.uk/en/Pub licationsandstatistics/Publications/PublicationsPolicyAndGuidance/DH_4119710, downloaded 17th December 2007.

DoH, Department of Health. (2007a) 'Performance Assessment Framework of Strategic Health Authorities, The Framework and Assessment Questionnaire 2003/2004', available at http://www. dh.gov.uk/assetRoot/04/08/56/56/04085656.pdf, accessed 23rd August 2007.

DoH, Department of Health. (2007b) 'Estates Return Information Collection (ERIC)', Webpage dated 1/5/07 available at http://www.dh.gov.uk/en/PolicyAndGuidance/OrganisationPolicy/EstatesAndFaci litiesManagement/PropertyManagement/DH_4117912, downloaded 17th December 2007.

DoH, Department of Health. (2007c) 'AEDET Evolution Toolkit', available at http://design.dh.gov. uk/content/connections/aedet_evolution.asp

Friedriksen, O. (2002). 'Benchmarking Facility Management of Hospitals Using the EFFOMETER Method', 17th Congress of International Federation of Hospital Engineering, Bergen, pp. 171–180.

Gelnay, B. (2002) 'Facilities Management and the Design of Victoria Public Hospitals', Proceedings of the COB Working Commission on Facilities Management and Maintenance Global Symposium, Glasgow, pp. 525–545.

HEFMA, Health Facilities Management Association. (1999) 'The National Performance Framework', available at http://www.hefma.org.uk, retrieved 17th December 2007.

HEFMA, Health Facilities Management Association. (2001)'ERIC: Performance Management', available at http://www.hefma.org.uk, retrieved 17th December 2007.

HEFMA, Health Facilities Management Association. (2003)'Healthcare FM – An Evidence Based Approach', available at http://www.hefma.org.uk, retrieved 17th December 2007.

Kam-Shim, M. W. Y. (1999) 'Managing Change: Facilities Management at the Pamela Youde Nethersole Eastern Hospital', Facilities, Vol. 17, Nos. 3–4, pp. 86–90.

Kaplan, R. S. and Norton, D. P. (1991) 'The Balanced Scorecard – Measures that Drive Performance', Harvard Business Review, Jan/Feb, pp. 71–79.

Kocakulah, M. C. and Austill, A. D. (2007) 'Balanced Scorecard Application in the Health Care Industry: A Case Study', Journal of Health Care Finance, Vol. 34, No. 1, pp. 72–100.

Lavy, S. and Shohet, I. M. (2004) 'Integrated Maintenance Management of Hospital Buildings: A Case Study', Construction Management and Economics, Jan, Vol. 22, No. 1, pp. 25–34.

Liyanage, C. and Egbu, C. (2005) 'Controlling Healthcare Associated Infections (HAI) and the Role of Facilities Management in Achieving 'Quality' in Healthcare: A Three-Dimensional View', Facilities, Vol. 23, No. 5/6, pp. 263–277.

Liyanage, C. and Egbu, C. (2006) 'The Integration of Key Players in the Control of Healthcare Associated Infections in Different Types of Domestic Services', Journal of Facilities Management, Vol. 4, No. 4, pp. 245–261.

Maslow, H. (1948) 'A Theory of Motivation', Psychological Review, Vol. 50, pp. 370–398.

Mc Dougall, G., Kelly, J., Hinks, J., and Bitici, U. (2002) 'A Review of the Leading Performance Tools for Assessing Buildings', Journal of Facilities Management, Vol. 1, No. 2, pp. 142–153.

Miller, L. and May, D. (2006) 'Patient Choice in the NHS How Critical Are Facilities Services in Influencing Patient Choice?', Facilities, Vol. 24, Nos. 9–10, pp. 354–364.

Neely, A. (2005) 'The Evolution of Performance Measurement Research. Developments in the Last Decade and a Research Agenda for the Next', International Journal of Operations and Production Management, Vol. 25, No. 12, pp. 1264–1277.

Neely, A. and Al Najjar, M. (2006) 'Management Learning Not Management Control: The True Role of Performance Measurement', California Management Review, Vol. 48, pp. 99–114.

Neely, A., Gregory, M., and Platts, K. (2005) 'Performance Measurement System Design a Literature Review and Research Agenda', International Journal of Operations and Productivity Management, Vol. 25, No. 12, pp. 1228–1263.

NHS Estates. (2008) '*ASPECT Staff and Patient Environmental Calibration Toolkit*', available at http://design.dh.gov.uk/downloads/aspect/ASPECT_documentation_v010705.pdf, accessed 22nd May 2008.

NHS Executive. (1999) '*Public Private Partnerships in the National Health Service: The Private Finance Initiative*', Treasury Task Force Publications, Treasury Public Enquiry Unit, London, available at http://www.dh.gov.uk/asset/root/04/02/11/27/04021127.pdf

NHS National Patient Safety Agency. (2006) '*Patient Environment Action Team Assessments 2006*', available at http://patientexperience.nhsestates.gov.uk/clean_hospitals/ch_content/home/background.asp, retrieved 17th December 2007.

NHS National Patient Safety Agency. (2007) '*Better Hospital Food*', Webpage describing PEAT initiative available at http://195.92.246.148/nhsestates/better_hospital_food/bhf_content/introduction/home.asp, retrieved 17th December 2007.

Payne, T. and Rees, D. (1999) '*NHS Facilities Management: A Prescription for Change*', *Facilities*, Vol. 17, Nos. 7/8, pp. 217–221.

Preisser, W. F. E. and Vischer, J. C. (2004) 'The Evolution of Building Performance. An Introduction' in '*Assessing Building Performance*', Eds W. F. E. Preisser and J. C. Vischer, Elsevier, Oxford, pp. 3–15.

Pullen, S., Atkinson, D., and Tucker, S. (2000) '*Improvements in Benchmarking the Asset Management of Medical Facilities*', Proceedings of the International Symposium on Facilities Management and Maintenance, Brisbane, pp. 265–271.

Rangone, A. (1997) 'Linking Organisational Effectiveness, Key Success Factors and Performance Measures: An Analytical Framework', *Management Accounting Research*, Vol. 8, No. 2, pp. 207–219.

Roberts, R. R., Frutus, P. W., Ciavarella, G., Gussow, L. M., Mensah, E. K., Kampe, L. M., Straus, H. E., Gnanaraj, J., and Rydman, R. J. (1999) 'Distribution of Variable Costs versus Fixed Costs of Hospital Care', *Journal of American Medical Association*, Vol. 281, pp. 644–649.

Shaw, C. (2003) '*How can Hospital Performance be Measured and Monitored?*' WHO Regional Office for Europe's Health Evidence Network (HEN), WHO Regional Office for Europe, available at http://www.euro.who.int/Document/E82975.pdf, retrieved 12th December 2007.

Then, D. S. S. (2004) 'Phase 6: Adaptive Reuse/Recycling – Market Needs Assessment' in '*Assessing Building Performance*' Eds W. F. E. Preisser and J. C. Vischer, Elsevier, Oxford, pp. 80–89.

WHO, Regional Office for Europe. (2003a) '*Measuring Hospital Performance to Improve the Quality of Care in Europe: A Need for Clarifying the Concepts and Defining the Main Dimensions*', Report on the 3rd and 4th WHO Workshop, Barcelona, Spain, January, available at http://www.euro.who.int/Document/E78873.pdf, retrieved 12th December 2007.

WHO, Regional Office for Europe. (2003b) '*Selection of Indicators for Hospital Performance Measurement*', Report on the 3rd and 4th WHO Workshop, Barcelona, Spain, June and September, available at http://www.euro.who.int/Document/E84679.pdf, retrieved 12th December 2007.

Hard Facilities and Performance Management in Hospitals

15

Igal M. Shohet and Sarel Lavy

This chapter reviews the state of the art in the domain of healthcare facilities management (FM). The review is followed by the introduction of an integrated key performance indicator (KPI) model for hospital maintenance. A case study is presented to illustrate the implementation of the KPIs in a hospital setting. The discussion concludes with a vision of a maintenance performance toolkit for healthcare FM.

15.1 Components of healthcare facilities management

Maintenance management of hospital buildings is one of the more complex subjects in the field of FM. Contributing to this are the great complexity of healthcare facilities, the high criticality of mechanical and electrical systems and the scarcity of maintenance resources. Furthermore, the performance and operation of hospital buildings are affected by numerous other factors, including actual hospital occupancy relative to planned occupancy, age of buildings, building environment, managerial resources invested and labour sources for implementing maintenance (in-house provision vs. outsourcing) (El-Haram and Horner, 2002; Holmes and Droop, 1982; Son and Yuen, 1993).

The current state of the art in healthcare FM, in both academic and professional communities, indicates the following domains as core disciplines in the area of healthcare FM: maintenance management, performance management, risk management, supply services management and development, along with information and communication technology (ICT) as an integrator (Shohet and Lavy, 2004). The following sections review these topics, and then introduce an integrated approach to hospital facilities maintenance.

15.1.1 Maintenance management

Maintenance management is one of the core domains of knowledge around which FM revolves. It includes not only the budgeting and priority setting of the different maintenance activities according to the preferred maintenance policy, but also service life planning. In order to achieve an optimal balance between cost minimisation and performance maximisation, the maintenance of complex facilities can be resolved using one of two alternatives: maximisation of performance level while maintaining a limited maintenance budget, or minimisation of costs subject to a minimum required performance level of the building (Jardine et al., 1997). Three maintenance policies can be implemented in built facilities: preventive maintenance, condition-based maintenance and breakdown maintenance (Tsang, 1995; Wang and Christer, 2000). Another strategic issue that affects the effectiveness and efficiency of maintenance is the allocation of resources in-house versus outsourcing. Outsourcing of maintenance entails the selection of appropriate procurement methods—performance-based or prescriptive-based procurement and specifications. Performance-based specification and procurement is based on the strength of performance specifications and focuses the procurement of maintenance along guidelines of performance of the built facilities. Prescriptive-based specification and procurement sets up the outline of the maintenance policy around prescribed specifications that impose adherence to specified methods rather than adherence to performance. The first concept allows for flexible implementation of maintenance along performance guidelines, whereas prescriptive specifications limit the flexibility of using predetermined methods of maintenance.

15.1.2 Performance management

Performance monitoring and management must be based on quantitative measures that characterise a facility's systems. This may also assist in the benchmarking process by which the facility's performance can be compared to that of other facilities, and thus identify the strengths and weaknesses of each facility. This procedure requires the identification, characterisation and definition of several KPIs, which are suitable for either public or private facilities (Alexander, 1996; Williams, 2000). These indicators may also be used as benchmarks for the cost-effectiveness of performance. The current state of the art indicates a need to develop integrated KPIs for healthcare facilities, seeking links between performance, maintenance, operations and energy expenditure and cost-effectiveness (Pullen et al., 2000).

15.1.3 Risk management

Risk management has gradually become one of the core themes faced by healthcare facility managers. In hospitals, different building systems and components, such as medical gases, fire protection systems and electricity, must exhibit high performance levels because any minor breakdown may lead to both casualties and financial losses. The current trend of cutting maintenance budgets affects risk levels by increasing the related risks, and hence forces facility managers to spend an increasing proportion of their time solving risk management problems (Okoroh *et al.*, 2002; Holt *et al.*, 2000). Risk management can be introduced into FM at the operational and strategic levels, using value engineering and value management.

15.1.4 Supply services management

Supply services management was discussed previously in reference to maintenance management. The topic, however, has an even broader aspect. It was previously noted that when dealing with maintenance and non-core activities, facility managers must find the appropriate procurement of maintenance for use of in-house and outsourced staff. Furthermore, supply services management also means determining the best combination of 'soft' FM services, such as cleaning, security, gardening, catering and laundry. FM is therefore required to find the best procurement arrangements for monitoring and analysing outsourced performance, and to assimilate the change through organisational learning.

15.1.5 Development

This domain encompasses a broad range of subjects pertaining to the mid-and long-range development of a facility. It includes strategic long-term planning, upgrading of existing facilities, rehabilitation, renovation, remodelling and reconstruction. Development is widely discussed in the literature on the subject; however, further research on the correspondence between the development of healthcare needs and supporting facilities is required (Jensen, 2008; Rondeau *et al.*, 2006).

15.1.6 Information and communication technology

One of the most important aspects of FM is information technology. Facility managers today are required to analyse all kinds of reports and rapidly deduce the steps to implement next. This emphasises the increased need and interest in the development of ICT applications for the domain of healthcare FM. Recognising different phenomena related to maintenance and operations problems is of great importance in the understanding of FM. Moreover, the complexities involved in the different facilities management themes, and their interrelations, can be solved and better understood if ICT is implemented (Yu *et al.*, 1997; Waring and Wainwright, 2002).

15.1.7 Summary

This review of literature discusses the core domains of healthcare FM. Gallagher (1998), for instance, defines the following six issues as encouraging successful implementation of healthcare FM: strategic planning, customer care, market testing, benchmarking, environmental management and staff development. Amaratunga *et al.* (2002) developed a model for assessing the impact of the organisation's FM cultural processes (SPICE-FM) on a hospital facility and concluded with a definition of the following issues as key processes for successful implementation of FM: service requirements management, service planning, service performance monitoring, supplier and contractor management, health and safety processes, risk management and service coordination. The authors of this chapter have identified the following six core domains within the area of healthcare FM: maintenance management, performance management, risk management, supply services management, development and ICT as an integrator between all other domains.

15.2 Key performance indicators in hospital facilities

Hospital facilities are normally composed of large and complex buildings, involving large-scale electrical and mechanical services and devices such as sophisticated backup systems, complex infrastructure and considerable medical systems, which are all unique to these facilities. Analytical tools are required for the quantification of the effects of occupancy and age on the performance and maintenance efficiency of these facilities.

Several KPIs have been developed for hospital buildings, and are discussed in the literature. Pullen *et al.* (2000) present seven KPIs for hospital facilities, most of which deal with business and financial performance. Lennerts *et al.* (2003, 2005) develop and discuss the processes involved in healthcare facilities management in Germany for reducing costs and increasing efficiency. The following paragraphs present seven major KPIs developed for use in healthcare facilities (Shohet, 2006), classified into four categories:

1. Asset development
2. Performance management
3. Maintenance
4. Organisational structure.

15.2.1 Asset development

Asset development KPIs conceptualise the effect of age and prevailing service conditions on the maintenance and performance of built facilities in order to develop an adjustment coefficient for the age of the facility, based on life cycle cost (LCC) concepts. These factors are described in the following text.

The facility coefficient is an adjusting coefficient that enables in-depth study of maintenance needs so as to adjust the allocation of resources to the actual service conditions of a particular building. This coefficient is computed using the following parameters: age of building, its physical environment (marine or in-land), average level of occupancy and configuration of building systems and components. As a result, it changes from one year to the next, even for the same building. Simulations of the values of this coefficient were carried out for

six typical combinations involving marine or inland environments, combined
with low, standard or high occupancies (see Table 15.1). Table 15.1 represents
the predicted values of the facility coefficient at 5-year intervals along the service life of a hospital building, for each of the six combinations studied between
the category of environment (2-point scale) and the level of occupancy (3-point
scale). The Facility Coefficient expresses the ratio between the maintenance
resources required for implementing comprehensive maintenance activities, and
the average resources required for implementing the same maintenance policy
in an inland hospitalisation building at standard occupancy over a service life of
75 years. This indicator can be used by facility managers to plan the allocation
of resources for the maintenance of facilities and buildings. It can also be used
for short- versus long-term planning of maintenance, and to determine the LCC
of a given building for any given year of its service life. The comprehensive relationship among service conditions, maintenance and performance is discussed in
a separate chapter. The reader may refer to another publication (Lavy and Shohet,
2007a) that discusses the analytical development of this indicator.

Table 15.1 Results of facility coefficient simulations.

Actual service life	Inland environment			Marine environment		
	Low occupancy	Standard occupancy	High occupancy	Low occupancy	Standard occupancy	High occupancy
5	0.37	0.42	0.51	0.40	0.45	0.54
10	0.45	0.53	0.66	0.47	0.56	0.68
15	0.83	0.86	1.03	0.85	0.88	1.06
20	1.14	1.20	1.49	1.21	1.26	1.55
25	1.07	1.20	1.36	1.16	1.29	1.45
30	1.03	1.08	1.11	1.07	1.12	1.15
35	1.27	1.35	1.60	1.26	1.34	1.58
40	1.52	1.53	1.79	1.54	1.55	1.80
45	1.19	1.40	1.50	1.25	1.46	1.57
50	0.95	1.30	1.21	0.99	1.34	1.26
55	1.19	1.23	1.29	1.22	1.26	1.32
60	1.16	1.23	1.60	1.20	1.28	1.64
65	0.70	0.84	1.05	0.74	0.88	1.09
70	0.41	0.42	0.51	0.45	0.46	0.55
75	0.40	0.39	0.50	0.43	0.43	0.53
Total	68.29	75.00	86.06	71.14	77.86	88.92

15.2.2 Performance management

The building performance indicator (BPI) monitors the physical condition and fitness for use of a building and its various systems and components. With this indicator, each building system receives a score on a scale ranging from 0 to 100 points. This parameter expresses the building's physical functional performance. For each building, the following ten systems are examined: structure, exterior envelope, interior finishes, electricity, heating, ventilation and air-conditioning (HVAC), plumbing, fire protection, elevators, communication and low voltage and medical gases. The overall score for a building is calculated by multiplying each system's score by its share in the building's total LCC, as shown in Eqn 15.1.

$$\text{BPI} = \sum_{n=1}^{10} P_n \times W_n \qquad (15.1)$$

where BPI is the building performance indicator; n, the index of building systems; P_n, the physical performance score of system n; and W_n, the weight of system n in the total LCC of the building. For hospital buildings, a BPI higher than 80 points indicates that the building is in good condition whereas a BPI lower than 60 points indicates a poor/dangerous level of performance for the building. This economic performance indicator allows the facility manager to obtain a comprehensive overview of the level of performance for each building and express it in an economic context. In addition, it provides insight into the physical performance of the building's systems and components.

15.2.3 Maintenance

The annual maintenance expenditure (AME) is an indicator used to evaluate the actual expenditure incurred by the maintenance department. It is calculated by adding up the annual expenditure on all sources of both in-house and outsourcing labour, as well as the costs of materials and spare parts. The AME is measured in units of $US per square metre.

The normalised annual maintenance expenditure (NAME) is based on the AME. The NAME expresses the annual maintenance expenditure neutralised for the effects of age, occupancy and category of environment. It provides a clear and transparent perspective on the AME after considering the facility coefficient, as shown in Eqn 15.2.

$$\text{NAME} = \frac{\text{AME}}{\text{FC}_y} \qquad (15.2)$$

where NAME is the normalised annual maintenance expenditure; AME, the annual maintenance expenditure and FC_y, the facility coefficient for year y.

The maintenance efficiency indicator (MEI) expresses the efficiency with which maintenance is implemented. It takes into account several factors including the annual maintenance expenditure, the building performance indicator and the facility coefficient, as shown in Eqn 15.3. The MEI therefore reflects the level of expenditure with reference to actual performance, the age of the hospital, its occupancy level and its environment.

$$MEI = \frac{AME}{BPI \times FC_y} = \frac{NAME}{BPI} \qquad (15.3)$$

where MEI is the maintenance efficiency indicator; AME, the annual maintenance expenditure; BPI, the building performance indicator; FC_y, the facility coefficient for year y and NAME, the normalised annual maintenance expenditure. The range of MEI values highly depends on the type of building and country in which it is located. For example, three ranges for MEI were established for hospital facilities in Israel (2008 values):

1. lower than 0.50, representing high efficiency of maintenance resource utilisation and/or lack of resources;
2. values between 0.50 and 0.70 indicating normative use of maintenance resources, hence being the desirable range of values for hospital facilities in Israel and
3. higher than 0.70 indicating high inputs compared with actual performance and/or a surplus of resources.

As this study was conducted on Israeli large acute-care hospital facilities, the MEI values represent the industry's average construction costs and best maintenance practices, in terms of policy-setting and procurement methods. Furthermore, the facility coefficient may represent prevailing conditions typical of the Israeli health sector, such as a relatively high occupancy compared with international standards and the proximity of large hospital facilities to the coastline. As these conditions and figures may vary between countries, the range of MEI may also vary; however, the method of its analysis is robust. Particular values for different countries can be adapted and fine-tuned.

Facility managers in hospitals can use the MEI for a comprehensive performance-based evaluation of the cost-effectiveness of maintenance operations. High MEI values indicate that the physical performance is lower than expected, based on the given maintenance resources; lower MEI values represent efficient use of resources, although they may also indicate a shortage of maintenance resources, which should be addressed by decision-makers in the individual facility. This indicator may also be used as a tool by higher-level decision- and policy-makers in allocating maintenance resources among several facilities.

15.2.4 Organisation and management

The manpower sources diagram (MSD) expresses the composition of maintenance personnel in terms of in-house staff versus outsourced staff. Shohet *et al.* (2003) distinguishes between two scenarios:

1. High-occupancy hospital facilities justify the use of in-house personnel due to accelerated deterioration of some interior and electromechanical building systems under intensive service conditions. This means that high availability of maintenance workers is required, and therefore, under such conditions, the employment of in-house personnel offers potential savings.
2. Low- and standard-occupancy hospital facilities entail potential savings by establishing a personnel composition, whereby the majority of maintenance workers are external employees.

This indicator is calculated as a percentage of the total outsourcing expenditure out of the total annual maintenance expenditure (TAME). The MSD indicates the composition of labour and the conditions under which in-house or outsourced resources should be preferred.

The managerial span of control (MSC) is one of the key managerial parameters that indicate the effectiveness of an organisation in achieving coherence among its units (Mintzberg, 1989). This indicator is defined as the ratio between the number of managers and the respective number of personnel directly subordinate to them. Although a wide span of control may save overhead expenses, it sometimes creates difficulties and ineffectiveness in control. However, a narrow span of control may save the amount of day-to-day coordination required and improve planning and control, but overhead expenses are eventually high. Even though this is not considered as a traditional hard FM indicator, the conclusions drawn from this indicator for the core of hard FM KPIs have necessitated the use of this indicator to establish effective procurement schemes. The MSC may support any required organisational changes in the FM department with the objective of improving its efficiency. It should therefore be used simultaneously with other KPIs.

15.3 Research methods

The following methods were used in this research: structured field survey (data gathering), statistical analysis and inference, model development and computing, validation of the model and sensitivity analyses.

15.3.1 Structured field survey

The field survey was carried out by using a structured questionnaire, aimed at identifying core parameters for the management of healthcare facilities. The survey focused on maintenance and operational aspects throughout the service life of the facilities. The questionnaire was distributed to healthcare facility managers, and responses were collected from 19 public acute-care hospitals in Israel. The facility managers were asked to provide information regarding core parameters of different aspects of FM in the hospital for which they are responsible, including characteristics of the hospital facility, building components and capacities, maintenance management, physical performance, energy consumption and operations.

15.3.2 Statistical analysis

The data collected in the field survey were analysed using descriptive statistics (bar histograms, pie charts, mean values and standard deviations), as well as inferential statistical methods (correlation coefficients, regression analysis, significance tests and t-tests). The descriptive and inferential statistical analyses revealed the main parameters that affect healthcare FM. These parameters led to the development of preliminary indicators for the evaluation of maintenance, performance, risk and management of healthcare facilities.

15.3.3 Model development and computing

The model development outlines five core topics healthcare FM confronts: maintenance, performance and risk, energy and operations, business management and development. These aspects, therefore, were suggested as the five core modules integrated into the developed model. The core topics, as well as the parameters included in these five modules, are as follows:

- Maintenance management module includes the profile of the facility and its maintenance in categories such as maintenance expenditure, distribution of labour resources, maintenance policies, maintenance efficiency, actual age of buildings and the level of occupancy.
- Performance and risk management module deals with actual, required and predicted physical performance levels for the facility, and the different buildings, systems and components in it.
- Energy and operations module refers to the energy consumption (water, electricity, fuel and medical gases). It also deals with the preparedness of the facility, defined as the capacity of energy reservoirs (water, fuel, power supply, fire extinguishing and air-conditioning), security, cleaning and gardening.
- Business management module includes the business aspects involved in the field of healthcare facilities management, such as administrative issues, taxation, insurance, and so on.
- Development module covers a broad range of subjects pertaining to the mid- and long-term development of a facility. It includes strategic long-term planning, upgrading of existing facilities, rehabilitation, renovation, remodelling and reconstruction.

The conceptual development of the model, developed in the frame of this research, encompassed the establishment, definition and formalisation of 15 procedures in the first two modules: maintenance management, and performance and risk management.

Microsoft Visual Studio.Net 2003© computer language was used to create a computer application in which the model can be applied. This consists of three interfaces: (1) an input interface, in which the user is required to provide a variety of data concerning the maintenance, performance and risk aspects of the facility; (2) reasoning evaluator and predictor phase, in which the different procedures developed during the theoretical and conceptual development of the model are implemented, and their results calculated and analysed and (3) an output interface, in which the main results are presented. These interfaces are discussed and presented in detail in Lavy and Shohet (2007b).

15.3.4 Validation
Several case studies were carried out in different acute-care hospitals in Israel, in which the model was evaluated and validated, and, as a result, improved and refined. The results obtained from these hospitals in a 2004 field survey were compared with those from 1999 and 2001 field surveys. Moreover, the validity of the model was examined by analysing and comparing the 2004 data with the predictions made for 2004, as a result of the original 1999 and 2001 surveys. The comparison of the results revealed several issues, in which refinement of the model was required and was therefore carried out. Furthermore, the validation process included conducting several sensitivity analyses of the model's results for: (1) possible inaccuracies in the evaluation of performance scores; and (2) the basic hypotheses of a linear pattern of deterioration for some building components.

15.4 Analysis of a hospital using the indicators developed – a case study

The developed indicators were implemented on several case studies. One of the more interesting cases is presented in detail in the following paragraphs, which review the background, findings and analysis of the results obtained for a hospital facility.

15.4.1 Profile of the hospital
Table 15.2 presents the hospital parameters for this case study. Based on the total floor area of the hospital in question and the number of patient beds, the average occupancy of this hospital was found to be 9.27 beds per 1000 sq. m of floor area. The reference number used in this study was taken from the Israeli Ministry of Health guidelines, which refer to a designed standard occupancy level

Table 15.2 Summary of hospital characteristics (1999 values).

Variable	Value
Built floor area (sq. m)	114,880
Number of patient beds	1,065
Average occupancy (number of beds per 1000 sq. m)	9.27
Average age of the buildings (years)	23
Annual outsourcing expenditure (10^3 $US)	2,205 (49.9%)
Annual in-house personnel expenditure (10^3 $US)	1,740 (39.4%)
Annual materials expenditure (10^3 $US)	475 (10.7%)
Total annual maintenance expenditure (10^3 $US)	4,420 (100.0%)
Annual maintenance expenditure ($US per sq. m)	38.4
Annual maintenance expenditure per patient bed ($US)	4,150
Average annual maintenance expenditure (% of reinstatement value)	2.29
Total number of in-house maintenance employees	58
Managerial span of control – Principal Engineer level	7

in hospital buildings of 10 patient beds per 1000 sq. m floor area. This means that the hospital was occupied at a level slightly lower than standard, or at almost 93% of standard occupancy. The AME was found to be US $52.1 per sq. m (2008 values), which is equivalent to US $5620 per patient bed. The distribution of this expenditure shows that almost half (49.9%) of it was spent on outsourcing providers of maintenance services, whereas 39.4% was allocated to internal, in-house, maintenance personnel. The rest of the expenditure (10.7%) was used to acquire materials and spare parts.

Regarding in-house maintenance personnel, it was found that 58 internal employees were engaged in hospital maintenance. This represents a ratio of 1980 sq. m floor area per in-house maintenance employee, or, alternatively, 18.4 patient beds per in-house maintenance employee. This ratio is more than 20% higher than the average ratio, as calculated for the sample population, which was found to be 1560 sq. m floor area per in-house employee, or 12.9 patient beds per employee. As the sample population comprises hospitals characterised by normative efficiency of maintenance implementation, the average ratio of patient beds per in-house maintenance employee represents a normative ratio. The formal managerial span of control (MSC) at the Principal Engineer level was three (cost accountant, quality assurance officer and maintenance manager). However, the maintenance manager's position was unoccupied so that the actual MSC was in fact seven (cost accountant, quality assurance officer, electrical engineer, air-conditioning engineer, water and sanitary systems engineer and civil engineer), taking into account all subordinates of any potential maintenance manager.

Table 15.3 Distribution of in-house maintenance employees by trade.

Maintenance profession	Number of employees
Electricity	16
Water and plumbing	11
Air-conditioning	7
Metal workshop	6
Carpentry workshop	5
Medical gases	3
Other (Principal Engineer, engineers and staff)	10
Total number of in-house maintenance employees	58

The distribution of in-house maintenance employees by profession was also investigated, as presented in Table 15.3. The professional profile of the maintenance employees reveals the use of very large teams in the electricity (16) and water and plumbing (11) trades. However, small teams of construction workers were used in other building trades, such as metal workshop (6) and carpentry (5).

All these data were gathered simultaneously with a survey of the physical condition of various building systems and components. Table 15.4 presents a summary of the building performance survey and the resultant BPI. Based on the findings of the buildings performance survey, it is evident that the entire facility was in deteriorating condition, which was perceived to become even worse if a drastic maintenance plan was not implemented immediately. The main systems that exhibited poor performance and maintenance levels were low voltage and communication, interior finishes, HVAC, electricity and exterior envelope. Only three building systems were found to be in satisfactory or good condition: fire protection, medical gases and elevators.

15.4.2 Data analysis

As shown in Table 15.4, the BPI in the above-described case study indicates that the buildings were in a deteriorating state (BPI = 66.1). The facility coefficient, which represents the age of the buildings, their occupancy level and surrounding environment, was calculated as 1.264. This value means that the hospital reached a point at which the annual maintenance expenditure is expected to be 26.4% higher than that for standard conditions, as described previously. This is attributed to the age of the buildings (23 years). The AME of US $52.1 per sq. m, coupled with a facility coefficient of 1.264, led to an MEI value of 0.62. This value expresses reasonable use of maintenance resources. It also shows that the BPI for the hospital campus correctly reflects the annual level of resources allocated for maintenance. Despite the sufficient MEI value, the AME seems to be low, that is, an increase in resources allocated to maintenance is required to improve the deteriorating condition of the buildings, as reflected by the BPI.

Table 15.4 Building performance indicator (BPI).

Building System	Performance	Special Remarks
Fire protection	100.0	-
Medical gases	100.0	-
Elevators	82.9	Few failures caused by high loads on elevators
Sanitary systems	79.2	Few pipe leakages and clogging of sewerage pipelines
Structure	72.0	Non-preventive inspections, breakdown maintenance
Exterior envelope	67.7	Leakage from roof and exterior walls
Electricity	66.7	Power failures (6–12 times per year)
HVAC	65.0	Failures in tubing, equipment, and end devices caused by high loads on the system
Interior finishes	56.1	Floors and acoustic ceilings in poor condition, non-periodical inspections, breakdown maintenance
Low voltage and communications	25.0	Many failures in the patient–nurse calling system, non-periodical inspections, breakdown maintenance
Total BPI	66.1	

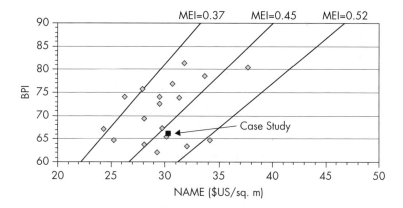

Figure 15.1 NAME versus BPI values.

Using these parameters, the NAME can be evaluated to be US $41.2 per sq. m. This value can be plotted against the BPI and compared with that of other hospital facilities, as depicted in Figure 15.1 (a benchmarking tool for measuring efficient hospital performance): the figure was created from a wide survey of public acute-care hospitals in Israel. The benchmarking reveals that the hospital examined in this case study utilises its resources at an acceptable level, yet its physical performance is relatively low compared with other hospitals. Improving the physical performance to at least 70 points, while retaining the current level of resource utilisation, would place this facility in a much better, yet still marginal position, whereby it would be comparable to other similar facilities. Hence, this was recommended for the development of tactical and strategic plans for this hospital.

Calculating the in-house versus outsourcing components of the total costs of maintenance shows that the costs of maintenance-related human resources were 44% from in-house provision versus 56% from outsourcing. The average occupancy of this hospital (lower than standard), together with its location in a metropolitan area in which availability of outsourcing is high, appears to justify the observed composition of labour. Moreover, the employee per square metre ratio for this facility, compared with the average ratio, may indicate a shortage of internal maintenance employees. Expansion of the internal maintenance department should, however, be carried out together with an increase in the expenditure for external contracting.

The actual MSC at the Principal Engineer level is seven, as long as the maintenance manager position is unoccupied. This represents a wide MSC at the Principal Engineer level. Although this situation seems to present some overhead savings, in this case it actually leads to difficulties in the Principal Engineer's ability to control. Instead of having a maintenance manager whose duties include decision-making, the Principal Engineer must confront these difficulties on a daily basis. Based on this, the projected amount of maintenance resources can be derived, as shown in Figure 15.2a. This figure presents the actual maintenance and performance condition in a two-dimensional manner; the horizontal axis represents the BPI, and the vertical axis represents the supplementary budget. The graph also represents the respective MEI levels. In order to enhance the building's performance, three alternatives were developed and considered, as shown in Figure 15.2b.

15.4.2.1 Alternative 1

Improving the BPI by increasing the efficiency level (MEI). This alternative does not require any additional investment; however, it supports the implementation of some organisational changes. The first step would be to hire a maintenance manager, reducing the Principal Engineer's span of control and perhaps increasing the department's effectiveness. Additional required organisational changes can be achieved by restructuring the composition of the various maintenance teams. Table 15.3 shows the distribution of in-house maintenance employees in the different building trades, whereas Table 15.4 shows the physical performance of these building systems. Cross-comparison of these two tables reveals that the electricity and communication team, for example, includes 16 workers, whereas the electrical system

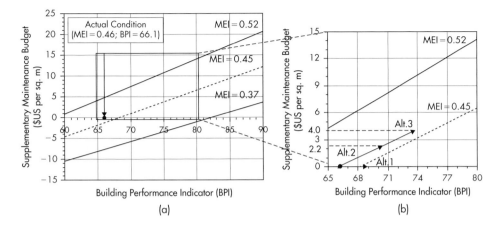

Figure 15.2 (a) Supplementary maintenance budget versus BPI for different levels of MEI; (b) three proposed alternatives for improving hospital performance.

is in deteriorated condition (66.7 points) and the communication and low-voltage system in poor condition (25.0 points). This leads to the conclusion that shifting labour proportions, that is minimising the in-house team and promoting greater use of outsourcing, may lead to better performance of this system. The HVAC system represents an opposite example, considering that only seven in-house maintenance workers throughout the entire hospital are charged with the responsibility for this system. The HVAC system was also found to be in a deteriorating condition (65.0 points). This leads to the conclusion that increasing the internal maintenance resources allocated for the HVAC system might improve its physical condition; thus, the recommendation would be to expand the in-house staff.

15.4.2.2 Alternative 2

Improving the BPI to 70 by preserving the actual level of efficiency (MEI = 0.62). This alternative will lead to enhanced building serviceability, despite the fact that its execution will not improve the deteriorated systems to a sufficient level of performance (70 points). As evident from Figure 15.2b, this alternative requires an additional capital expenditure of US $2.7 per sq. m of floor area, which can be divided into an additional US $2.4 per sq. m for labour and US $0.3 per sq. m for materials and spare parts. This results in a total supplement of US $310 000, which constitutes a 5.2% increase in the actual maintenance expenditure. Alternative 2 also implies organisational changes similar to those suggested for the implementation of Alternative 1. Figure 15.3 shows the different options of applying this alternative on an MSD for an addition of US $2.4 per sq. m for labour. Four major options can be implemented: Alternative 2A proposes retaining in-house labour as is, coupled with an increase in outsourcing resources. Alternative 2D, however, proposes equalising internal and external resources, which means an

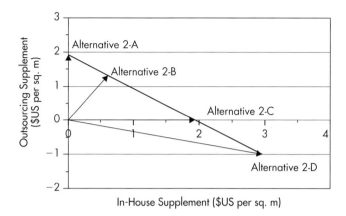

Figure 15.3 Manpower sources diagram (MSD) options for Alternative 2.

increase in in-house labour coupled with a decrease in outsourcing of maintenance resources. Alternatives 2B and 2C are two mid-way options between 2A and 2D, proposing either to maintain the current resource composition (Alternative 2B) or to increase the in-house labour component only (Alternative 2C). It is clear from Figure 15.3 that three of the four proposed alternatives involve an increase in internal sources (Alternatives 2B, 2C and 2D), whereas Alternative 2A is based on retaining their current level. Owing to the hospital's parameters (lower than standard occupancy, existing composition of human resources and the location of the hospital in the centre of a metropolitan area) the recommendation would be to continue to rely on external sources of labour, although the internal component may be slightly increased due to the low ratio of maintenance employees per built area compared with the average. This analysis should lead hospital decision-makers to prefer the implementation of Alternative 2B.

15.4.2.3 Alternative 3
Improving all building systems that are in deteriorating condition to a performance level of 70 points (i.e. achieving a total BPI greater than 70), while preserving the existing level of efficiency (MEI). As this alternative will enhance the overall hospital's performance, it requires additional financial and human resources. Its implementation will, therefore, require a capital expenditure of US $5.2 per sq. m (as seen in Figure 15.2b), which may be divided into US $4.6 per sq. m for labour and US $0.6 per sq. m for materials and spare parts, for a total of US $600 000. This represents a 10.0% increase over the actual maintenance expenditure, which is a significant supplement. Alternative 3 also includes the organisational changes suggested for the implementation of Alternative 1. The analysis of the different MSD options for Alternative 2 can be adapted to the examination of the human resources under Alternative 3.

15.4.3 Conclusions

The following conclusions can be drawn from the hospital case study analysis:

1. The MEI (0.62) shows that the BPI (66.1) is compatible with the existing annual level of available resources (US $52.1 per sq. m floor area or 2.29% of reinstatement value). This already takes into consideration the hospital's level of occupancy, the age of the buildings and the type of environment in which the hospital is located.

2. The BPI may be improved by one of the following three proposed alternatives:

   a. Alternative 1 focuses on organisational and labour improvements and does not require any additional investment in maintenance. Implementation of this alternative may improve the BPI by approximately 2–3 points, to a BPI level of 68–69 points, while improving the MEI to a level of 0.6.

   b. Alternative 2 focuses on organisational improvement, together with an increase in labour resources, and it requires an additional investment of US $310 000. Implementing this alternative may improve the BPI by approximately 4 points, to a BPI level of 70, while maintaining the current MEI.

   c. Alternative 3 also focuses on organisational improvement, together with a significant increase in labour resources, but it requires an additional investment of US $600 000. Its implementation may improve the BPI by approximately 7 points, to a BPI level of 73 points, while maintaining the actual level of MEI.

3. The scarcity of internal labour resources (one employee per 1980 per sq. m built floor area) could explain the low BPI level. Thus, an addition of internal employees is recommended in this case, especially for the following systems: exterior envelope, interior finishes and communications and low voltage. On the other hand, owing to the special characteristics of this hospital (low occupancy and accessibility to outsourcing), increasing the proportion of outsourcing in the total expenditure is also recommended, especially for the electricity and plumbing systems.

4. A higher level of maintenance efficiency may be achieved by hiring a maintenance manager. This would reduce the Principal Engineer's span of control and, consequently, would decrease the maintenance teams' dependence on the Principal Engineer and improve effectiveness.

Results expected from the implementation of these alternatives include improvement in the BPI, which can be obtained by carrying out activities in various building systems, such as communications, interior finishes, HVAC and exterior envelope. Such activities should lead to a decrease in the number of failures and include the administration of routine periodic inspections. Top-priority issues should, however, be determined by both the hospital board and the hospital's FM professionals.

15.5 Discussion

Hierarchical reasoning for tactical and strategic FM decision-making, as demonstrated in the case study discussed in the preceding text, is essential for successful healthcare FM. This reasoning mechanism implements integrated analyses of KPIs that shed light on the organisational effectiveness and efficiency of healthcare FM, and on performance and maintenance policy-setting.

Hospitals must operate 24 h a day, 7 days a week, providing emergency intensive and life-saving care and treatment services. They also encompass the critical infrastructure of healthcare, such as power supply for operating theatres, and medical gas in intensive care units. Therefore, when making decisions, facility managers must consider many factors and their effects on owners, clients, services and staff. The approach introduced in this chapter integrates physical performance, financial, personnel and organisational aspects to obtain a quantitative measure that evaluates the parameters affecting the execution of maintenance activities. It presents seven indicators: BPI, AME, FC_y, MEI, NAME, MSD and MSC. An integrated analytical process was demonstrated, involving data gathering, analysis of KPIs and drawing of conclusions and recommendations based on the analysis and diagnosis of the findings.

Using the developed KPIs, guidelines for strategic FM policy setting may be outlined for the methodological design and operation of facilities from a life cycle and performance perspective. The development of the analytical quantitative model may significantly contribute to a better understanding of healthcare FM, to the measuring of efficiency and to improving FM performance.

15.6 Towards a maintenance performance toolkit

The conceptual model developed in this chapter stresses that maintenance management of hospital facilities implement integrated decision-making tools that address the following key issues:

- Systematic monitoring of the physical functional performance of the facility
- Systematic records of maintenance expenditures
- Quantification of the facility's occupancy level in terms of maintenance resources
- Quantification of the effect of the facility's actual service life on maintenance activities and expenditures
- Monitoring of the efficiency of the actual managerial resources for FM.

Maintenance of hospital facilities may be realised using a toolkit that encompasses four of the core topics of healthcare FM, using IT as an integrating tool. A computerised toolkit is proposed that covers performance, maintenance, management

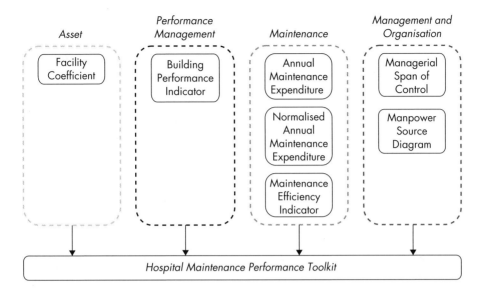

Figure 15.4 Hospital maintenance performance toolkit.

and organisation and asset development. The proposed toolkit enables monitoring of hospital facility performance, planning, control and management of maintenance, as shown in Figure 15.4.

References

Alexander, K. (1996). *Facilities Management: Theory and Practice*, E&FN Spon, London, U.K.

Amaratunga, D., Haigh, R., Sarshar, M. and Baldry, D. (2002). "Assessment of facilities management process capability: A NHS facilities case study", *International Journal of Health Care Quality Assurance*, 15(6), 277–288.

El-Haram, M. A. and Horner, M. W. (2002). "Factors affecting housing maintenance cost", *Journal of Quality in Maintenance Engineering*, 8(2), 115–123.

Gallagher, M. (1998). "Evolution of facilities management in the health care sector", *Construction Papers*, No. 86, 1–8, The Chartered Institute of Building, ed. P. Harlow.

Holt, B., Edkins, A. and Millan, G. (2000). "Developing a generic risk database for FM", in Nutt B. and McLennan, P. eds, *Facility Management – Risks and Opportunities*, Blackwell Science, Oxford, U.K., 201–211.

Holmes, R. and Droop, C. (1982). "Factors effecting maintenance costs in local authority housing", in Bradon, P. S. ed., *Building Cost Techniques: New Directions*, E&FN Spon, London, U.K., 398–409.

Jardine, A. K. S., Banjevic, D. and Makis, V. (1997). "Optimal replacement policy and the structure of software for condition-based maintenance", *Journal of Quality in Maintenance Engineering*, 3(2), 109–119.

Jensen, P. A. (2008), "The origin and constitution of facilities management as an integrated corporate function", *Facilities*, 26(13–14), 490–500.

Lennerts, K., Abel, J., Pfrunder, U. and Sharma, V. (2003), "Reducing health care costs through optimised facility management-related processes", *Journal of Facilities Management*, 2(2), 192–206.

Lennerts, K., Abel, J., Pfrunder, U. and Sharma, V. (2005), "Step-by-step process analysis for hospital facility management: An insight into the OPIK research project", *Facilities*, 23(3/4), 164–175.

Lavy, S. and Shohet, I. M. (2007a), "On the effect of service life conditions on the maintenance costs of healthcare facilities", *Construction Management and Economics*, 25(10), 1087–1098.

Lavy, S. and Shohet, I. M. (2007b), "Computer-aided healthcare facility management", *ASCE Journal of Computing in Civil Engineering*, 21(5), 363–372.

Mintzberg, H. (1989) *Mintzberg on Management – Inside Our Strange World of Organizations*, Free Press, New York.

Okoroh, M. I., Gombera, P. P. and Ilozor, B. D. (2002). "Managing FM (support services): Business risks in the healthcare sector", *Facilities*, 20(1/2), 41–51.

Pullen, S., Atkinson, D. and Tucker, S. (2000), "Improvements in benchmarking the asset management of medical facilities", *Proceedings of the International Symposium on Facilities Management and Maintenance*, Brisbane, Australia, pp. 265–271.

Rondeau, E. P., Brown, R. K. and Lapides, P. D. (2006), *Facility Management*, 2nd edn, John Wiley and Sons, Inc., Hoboken, New Jersey.

Shohet, I. M. (2006), "Key Performance Indicators for strategic healthcare facilities maintenance", *ASCE Journal of Construction Engineering and Management*, 132(4), 345–352.

Shohet, I. M. and Lavy, S. (2004). "Healthcare facilities management: State of the art review", *Facilities*, 22(7/8), 210–220.

Shohet, I. M., Lavy-Leibovich, S. and Bar-on, D. (2003) "Integrated maintenance monitoring of hospital buildings", *Construction Management and Economics*, 21(2), 219–228.

Son, L. H. and Yuen, G. C. S. (1993). *Building Maintenance Technology*, Macmillan Press Ltd., Hampshire, U.K.

Tsang A. H. C. (1995). "Condition-based maintenance: Tools and decision making", *Journal of Quality in Maintenance Engineering*, 1(3), 3–17.

Wang, W. and Christer, A. H. (2000). "Towards a general condition based maintenance model for a stochastic dynamic system", *Journal of the Operational Research Society*, 51(2), 145–155.

Waring, T. and Wainwright, D. (2002). "Enhancing clinical and management discourse in ICT implementation", *Journal of Management in Medicine*, 16(2/3), 133–149.

Williams, B. (2000). *An Introduction to Benchmarking Facilities and Justifying the Investment in Facilities*, Building Economics Bureau Ltd., Bromley, Kent, U.K.

Yu, K., Froese, T. and Vinet, B. (1997). "Facilities management core models", *Annual Conference of the Canadian Society for Civil Engineering*, Sherbrooke, Quebec, May 1997.

Community Clinics

Hard Facilities Management and Performance Management

Igal M. Shohet

Contemporary trends in healthcare services provision tend towards increased provision of healthcare services through community-based healthcare centres. This concept of healthcare provision is based on the hypothesis that the future hospital will be composed of the core healthcare services, such as intensive and acute care, operating theatres, whereas most of the primary and secondary care, such as diagnostics, rehabilitation and long-term care will be delivered through a network of community-based clinics. This architecture is developed on the conception that essential healthcare based on practical, scientifically sound and socially acceptable methods and technology made universally accessible to individuals and families in the community at a cost that the country can afford. This concept implies that a network of community clinics, equipped with state-of-the-art means for telemedicine, is to be established with a wide geographical dispersion. The aim of this chapter is to explore the core FM parameters in clinic facilities and to compare them with hospital facilities. This comparison provides an analytical background for a discussion on the contribution of community clinics as an effective infrastructure facility for primary and secondary healthcare delivery. This chapter reflects the results of research on healthcare facilities management (FM) carried out over the past 7 years. Seven key performance indicators (KPIs) were developed and implemented within the scope of this research. The KPIs delineate the performance and the effectiveness of maintenance services delivery in healthcare facilities. Comparing the performance and maintenance of clinic with those of hospital facilities reveals that the

249

economics and performance of clinic facilities entails a high potential for the accomplishment of improved services. This improvement may be accomplished through close community clinical services, higher physical performance of the built facilities and effective delivery of FM services.

16.1 Introduction

16.1.1 Healthcare facilities management

Many countries worldwide are witnessing similar trends in healthcare services provision. Triggered by the natural population growth, the ageing of the population and the consumer revolution, an increase in the demand for healthcare in public hospitals is observed (Hosking and Jarvis, 2003). Consequently, the total number of in- and out-patient admissions has increased as well. In order to deal effectively with the proliferation of in-patient admissions, and as a result of their limited resources, hospitals tend to minimise patients' average length of stay. This trend is evidenced in separate reports issued in the USA (AHA, 2006), Germany (Federal Statistical Office Germany, 2003) and the UK (Hensher and Edwards, 1999). In the USA, for example, an increase of more than 250% in the total in-patient admissions was observed between 1980 and 2004, a partial parallel matching of the resources enforced healthcare decision-makers to reduce the average length of stay by approximately 25% (AHA, 2006). These trends have led to a demand to investigate the structure of healthcare systems and FM decision-making processes in the industry. Melin and Granath (2004), for instance, presented a Swedish study that examined the effect of horizontal integrated care (HIC, which deals with ways in which care is delivered to patients) on healthcare FM and the implications of HIC, local hospitals and close care on the built environment. Rees (1997, 1998) examined the development of the FM profession within the National Health Services (NHS) in the UK, and concluded that NHS Trusts tend to integrate non-core services (e.g. risk management, energy efficiency, cleaning, security etc.) under the umbrella of the FM department. It was also observed that in only 24% of the Trusts, the senior FM director was a board-level executive.

Examination of FM in the healthcare sector reveals an underinvestment in the allocation of resources (AHA, 2006; British Ministry of Finance, 2003). This might adversely affect the non-core activities of healthcare providers, and primarily FM aspects, such as maintenance activities and operations. The American Hospital Association (AHA) states in its 2003 Annual Report that 'hospitals have been under financial pressure in the last five years, both from public and private payers. Since 1999, up to one third of hospitals have had negative total margins' (AHA, 2006). A similar state of affairs is presented in the 2003 Annual Report of the British Ministry of Finance, which states: 'Over the past 30 years, the UK has consistently invested a smaller share of its national income in healthcare than comparator countries. Historical underinvestment has resulted in poorer health outcomes than the EU average' (British Ministry of Finance, 2003).

Once every few years, the World Bank publishes a number of health indicators for each country in the world, and classifies them according to their average income level and regional categories. Two of the most interesting indicators are the health expenditure per capita index (in $US) and the total health expenditure as a percentage of the gross domestic product (GDP). The values of these two indicators in Israel rank relatively high compared with most countries in the world: the expenditure per capita is $1496, where the total health expenditure as a percentage of the GDP is 9.1%, lower by only 0.28% compared with the average for the European Monetary Union (World Bank, 2007).

In response to a steady demand to provide healthcare in more remote regions, the Israel healthcare system developed a network of clinics organised in a vertical hierarchical scheme. This network is composed of three levels, that is community clinics, hospitals and strategic hospitals (Figure 16.1): Community clinics that are located in every town or village, and are 500–2500 sq. m in size (with a mean size of 1200 sq. m). Eight hundred community clinics operate throughout the entire country, each providing primary care to an average of 8000 insurees. Community clinics are supported by a network of 40 regional clinics that provide secondary care services such as MRI, X-ray and medical consultancy, together with regional laboratories that supply diagnostic services to both community and regional clinics. This network acts as a screening mechanism that provides primary care to insurees of the Israeli system prior to their admission into a peripheral or regional hospital. Hospitals are classified into three categories according to infrastructures and medical care: peripheral hospitals (less than 400 patient beds), regional

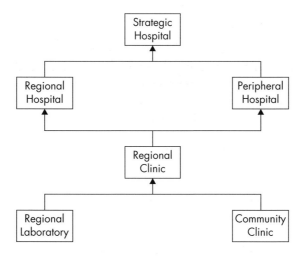

Figure 16.1 Alternative architecture of healthcare facilities.

hospitals (401–1000 patient beds) and strategic hospitals that provide unique specified care and that are equipped with built and advanced medical infrastructures (800–1500 patient beds).

All these trends and figures emphasise the necessity to expand the healthcare services with the development of wide distributed network of community-based clinics that will improve accessibility to healthcare and reduce the demand for hospital healthcare services. This objective can be realised through the establishment of a network of community clinics capable to effectively deliver primary healthcare services at distributed locations. Acute-care hospitals are witnessing an increased number of in-patient admissions, and at the same time, a decrease in the average length of stay. One possibility for bridging the gap between the growing demand for healthcare and limited resources of healthcare infrastructure is to develop a hierarchical network of community clinics designated to provide most of the primary care and part of the secondary care. Clinics may be classified into three categories: community clinic – providing preliminary diagnostic healthcare; regional clinics and laboratories – support community clinics with laboratory analysis instruments and providing secondary care such as rehabilitation – equipped with medical devices such as MRI, X-ray and telemedicine instruments.

16.1.2 Alternative architectures of healthcare service provision

The provision of healthcare services is traditionally delivered through a network of hospitals: regional and principal facilities. This concept is implemented in most Western European countries; it requires a large network of hospitals, which implies considerable capital investments and appropriate resources for maintenance of these facilities. According to this concept, the provision of close care in hospitals meets with difficulties and is nearly impossible, since hospital facilities are not readily accessible to populations such as the elderly and children, the costs of providing clinic facilities services is higher than costs of providing clinic facilities services in diverse sites at the community. Delivery of healthcare in community clinics provide accessible primary healthcare services at costs that the community and the country can afford. Furthermore, the capital reinstatement value of a hospital is $1790–$3300 per sq. m and the annual maintenance expenditure (AME) required to maintain such facilities is $35–54 per sq. m, depending on the level of performance provided (Shohet et al., 2003). Sixty percent of the resources required for maintenance are labour; thus, hospital maintenance departments are invariably labour intensive. Furthermore, such facilities are considerably sensitive to failures in critical systems such as power supply, medical gas, low voltage and communication.

An alternative architecture and infrastructure for the provision of primary care and part of the secondary healthcare is the provision of healthcare infrastructures in a vertical scheme: built facilities for ordinary primary close care that are located within the community and are equipped with infrastructures for primary care. Such clinics are supported by regional facilities that provide laboratories, radiology and diagnostics services (as outlined in Figure 16.1). Insurees can access a regional clinic only after visiting a community clinic and receiving a referral order to a regional clinic. This architecture is developed on the conception that essential healthcare based on practical, scientifically sound and socially acceptable methods and technology made universally accessible to individuals and families in the community at a cost that the community and the country can afford to maintain (WHO, 1978). Healthcare provided in remote large facilities such as hospitals is not necessarily readily accessible by some sectors of the community and may not meet these standards.

Healthcare in peripheral and regional hospitals is provided to insurees following the provision of primary care and diagnostics in regional clinics. This excludes emergency circumstances, in which insurees receive healthcare at emergency departments of acute-care hospitals. The following paragraphs delineate a profile of clinic facilities and discuss the potential of these facilities with respect to hospital facilities in terms of accessibility, performance of the facility for both the medical staff and patients, and the benefit-to-cost ratio in these facilities compared to hospital facilities; the implications of this profile on the economic and the performance of healthcare facilities as revealed by the analysis in the Israeli healthcare system are discussed.

16.2 Clinic facilities

A tool of KPIs, implemented and validated in various healthcare facilities, is presented here as an evaluation and assessment tool of healthcare facility management. The KPIs are used to analyse the core FM parameters of clinic and hospital facilities as a deduction tool.

16.2.1 Key performance indicators in clinic facilities

Seven KPIs for monitoring the performance and maintenance of clinic facilities were developed and implemented as follows. For each KPI, we present the hypotheses and theory underlying the core parameters that lay the ground for the subsequent analysis:

1. Age coefficient (AC_y).
2. Clinic's patient density coefficient (DC).
3. Building performance indicator (BPI).
4. Annual maintenance expenditure (AME).
5. Maintenance efficiency indicator (MEI).
6. Maintenance sources diagram (MSD).
7. Managerial span of control (MSC).

16.2.1.1 Research method

The research method of the development of the KPIs followed four stages:

- Gathering of data from clinic and hospital engineering departments. The data gathered included facility parameters (floor area, density of patients, age of facility, location), maintenance resources (labour, materials and outsourcing) and data on failures in building systems.
- The second stage of the research focused on a statistical comparative and quantitative analyses of maintenance expenditures under various service conditions: intensive and standard service conditions (Shohet *et al.*, 2002).
- Key performance indicators were developed based on the statistical and quantitative analyses.
- Application of the indicators to case studies so as to appraise the indicators developed.

16.2.1.2 Age coefficient

The AC adjusts a facility's maintenance needs to its actual service life. The hypothesis behind this coefficient is that the actual service life of building components affects the probability and consequently the rate of failure. Given a predicted life cycle and distribution of failure, the annual expenditure for replacements and preventive maintenance expenditures may be analysed and calculated. The ratio between the AME for a given year and the mean AME for any year is referred as the age coefficient and may be used as an adjusting coefficient for the effect of age on the annual expenditures on maintenance. Expression 16.1 calculates the AME for a given year by summing the annual expenditures predicted for preventive maintenance and the costs for replacements of each building system.

$$AME_y = \sum_{n=1}^{10} \left[\sum_{j=1}^{m} \left(M_{ynj} + R_{ynj} \right) \right] \tag{16.1}$$
$$\forall_y = 1, 2, 3, ..., 50$$

where, AME_y is the annual maintenance expenditure for year y; n is building systems counter; j is component index in system n; m is total number of components j in system n; M_{ynj} is maintenance costs of component j in system n for year y ($/sq. m); R_{ynj} is replacements costs of component j in system n for year y ($/sq. m); y is year counter for duration of the building service life.

Expression 16.2 calculates the mean AME along the life cycle of the building.

$$AME_{ave} = \frac{\sum_{y=1}^{50} AME_y}{50} \tag{16.2}$$

where, AME_{ave} is the average annual maintenance expenditure for the duration of the clinic service life (50 years) in dollar per square metre; AME_y is total annual maintenance expenditure for year y; y is year counter for duration of the building service life.

Expression 16.3 calculates a 10-year moving average of AC_y:

$$AC_y = \frac{\sum\limits_{y-4}^{y+4} AME_y + \frac{1}{2}\left(AME_{y-5} + AME_{y+5}\right)}{10 \times AME_{ave}} \qquad (16.3)$$

$$\forall_y = 6,7,8,...,50$$

where, AC_y is the age coefficient for year y; AME_y is total annual maintenance expenditure for year y ($/sq.m); AME_{ave} isaverage annual maintenance expenditure for the duration of the clinic service life ($/sq. m).

Simulation of the above coefficient for a typical clinic facility with a designed life cycle (DLC) of 50 years produced the following results (Figure 16.2): A value of 1 represents the average maintenance expenditure (2.5% of reinstatement value) for the duration of the clinic's designed life cycle. The total area below the graph equals 50. The graph depicts a maximum point at the middle of the facility DLC, indicating replacements of major components of electromechanical systems (e.g. electric boards, switch gears, HVAC mechanical and control units etc.), and multiple local maxima indicating clinic interior finishing renovation works.

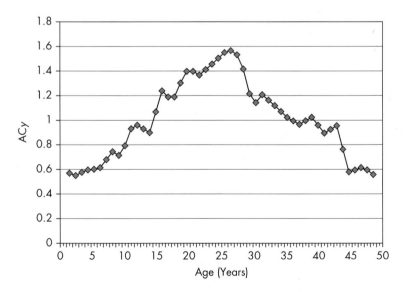

Figure 16.2 Age coefficient (AC_y) versus actual age of clinic for DLC of 50 years.

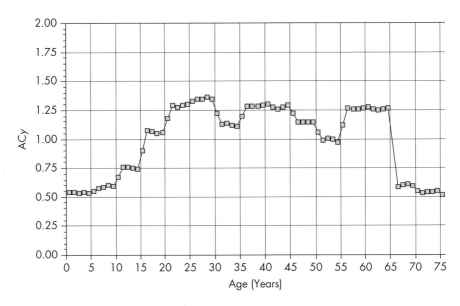

Figure 16.3 Age coefficient (AC$_y$) versus actual age of hospital for DLC of 75 years.

Figure 16.3 depicts a similar analysis carried out on hospital building. The graph delineates multiple local maxima and minima points that reflect replacements of major electro mechanical systems such as electric boards, heating and cooling systems, along the life cycle of the building.

16.2.1.3 Density coefficient

The DC quantifies the effect of patient density in the clinic on the clinic components' rate of deterioration. Standard density was defined as 175 patients per square metre per annum and is referred to as 100% patient density (Shohet and Kot, 2006; Shohet et al., 2008). The research hypothesis was that density conditions affect the deterioration pattern of building components and systems. The DC was developed based on the strength of analysis of the life cycle of building components under intensive and moderate service conditions. Results were as follows (Figure 16.4):

1. Under moderate-density conditions (less than or equal to 80% of the standard density), the DC equals 0.97, expressing only minor savings in maintenance activities because of a mandatory preventive policy with respect to systems such as electricity, fire protection, water and sanitation system.
2. When relative density is between 80% and 100%, the increase in maintenance activities is moderately linear, with a slope of 0.001625.
3. When relative density is between 100% and 154%, patient density has a greater impact on the maintenance expenditure, the slope of the graph increases to 0.00578, and the DC under high-density conditions remains constant at 1.31.

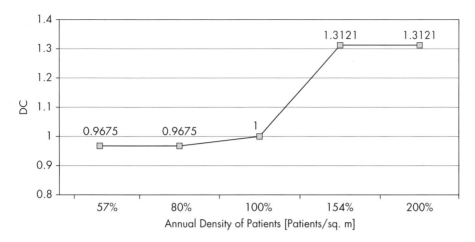

Figure 16.4 Density coefficient for clinic facilities.

16.2.1.4 Building performance indicator

This KPI enables to evaluate the overall state of a clinic or of a clinic portfolio, according to the performance of its components and systems. The indicator is defined by a value, ranging between 0 and 100, that expresses the clinic's performance, including that of its various systems (P_n). P_n is graded according to performance scales between 0 and 100, where $P_n < 60$ indicates poor/dangerous performance condition, $60 < P_n \leq 70$ indicates deteriorating performance condition, $70 < P_n \leq 80$ indicates marginal (70) or satisfactory (80) condition, and $P_n > 80$ indicates good condition of the clinic facility. The actual score for each system (P_n) is expressed by Eqn 16.4. It is composed of three components of facility maintenance: (1) actual condition of the system (C_n); (2) frequency of failures affecting the service provided by the system (F_n); and (3) actual preventive activities carried out on the system to maintain acceptable service level (PM_n) (Israel Standards Institution, 2002).

$$P_n = C_n \times W(C)_n + F_n \times W(F)_n + PM_n \times W(pm)_n \qquad (16.4)$$

where, $W(C)_n$ is the weight of component condition of system n; $W(F)_n$ is weight of failures in system n; $W(pm)_n$ is weight of preventive maintenance for system n.

For every system, the sum $W(C)_n + W(F)_n + W(pm)_n$ totals 1.

The score C_n is evaluated on the strength of a 100-point performance rating scale whereby a score of 100 expresses perfect performance, 60 represents – deteriorating performance and 40 and 20 represent failure and poor performance, respectively. The preventive maintenance score (PM_n) is evaluated on the basis of the maintenance policy governing the component and the frequency of proactive inspections carried out with respect to maintenance standards. Frequency of failures (F_n) is evaluated on a scale between 100 (no failure in 12 months) and

20 (frequent failures, for example for a roofing system, 12 times in the past 12 months). The combination of these three elements produces the performance score of the entire system (P_n). Weighting of each building system (W_n) in the BPI is accomplished by weighting the relative contributions of the system's components to the clinic's life cycle costs. Table 16.1 presents the weightings of clinic building systems. It is shown that the interior finishing system accounts for 24.4% of the clinic's life cycle costs, the structure accounts for 15.5%, the exterior envelope for 12.1 and the HVAC system for 10.5%. The profile of this breakdown emphasises that, for clinics, the interior finishing and electro-mechanical systems account for 63% of the entire BPI.

Once the systems' functional states have been diagnosed, the BPI is calculated. The BPI for each system is obtained by multiplying its weight by its score (Eqn 16.5).

$$BPI = \sum_{n=1}^{10} P_n \times W_n \qquad (16.5)$$

The desired BPI range is BPI>80, though at such a performance score, any individual system or component with a performance score below 70 requires corrective maintenance measures.

This parameter enables (1) to evaluate the overall state of a clinic; (2) to evaluate the state of the clinic's systems; (3) to benchmark the asset's performance in relation to other clinics or facilities (inter-organisational benchmarking); and (4) to benchmark the clinic's systems in order to compare the efficiency of the various maintenance crews (intra-organisational benchmarking).

Table 16.1 Breakdown of building performance indicator (BPI) into building systems (W_n) for clinic facilities.

Building system	BPI (W_n)
Structure	15.5
Exterior envelope	12.1
Interior finishing	24.4
Electrical systems	9.7
Water and waste water	5.8
HVAC	10.5
Fire protection	0.7
Communications	3.4
Peripheral infrastructure	9.6
Elevators	8.3
BPI	100.00

16.2.1.5 Annual maintenance expenditure

This parameter reflects the scope of expenditure per square metre built area (excluding cleaning, energy and security expenditures). From an organisational viewpoint, this parameter determines the annual expenditure on maintenance of the clinics, and it can also provide a measure of the overall expenditure on built assets in relation to the organisation's turnover. From a professional FM viewpoint, however, expenses must be analysed in relation to the clinic's characteristics and with respect to the output (the clinic's performance). This examination is carried out in the framework of the maintenance efficiency indicator (MEI), as described in the following section.

16.2.1.6 Maintenance efficiency indicator

This indicator enables to examine the investment in maintenance in relation to the clinic's performance (which is in fact the service the FM department provides to the healthcare organisation). The MEI is calculated using Eqn 16.6:

$$MEI = \frac{AME}{AC_y} \times \frac{1}{BPI} \times \frac{1}{DC} \times i_c \qquad (16.6)$$

where, AME is the actual annual maintenance expenditure; AC_y is age coefficient for year y; BPI is monitored building performance indicator; DC is density coefficient for the clinic in question; and i_c is construction prices index.

The MEI expresses the expenditure on maintenance per clinic performance unit, after neutralising the effects of age (AC_y) and patient density (DC).

The MEI can be analysed using a two-dimensional diagram of BPI (dependent variable) and the normalised annual maintenance expenditure (NAME) (independent variable), as expressed in Eqn 16.7:

$$NAME = \frac{AME}{AC_y \times DC} \qquad (16.7)$$

NAME expresses the annual maintenance expenditure after neutralising the effect of age (AC_y), and patients' density (DC) of the clinic.

For a clinic that is maintained at a desired level, we assume a BPI of 100. The average AME per square metre was found to be 2.50% of the reinstatement value of a clinic facility, which was calculated to be $1180 per sq. m built. Assuming a facility with an AC of 1.00 (the standard), and a DC of 1.00, the MEI value would be 0.30. The upper and the lower margins of the desirable range were deduced from the MEI's standard deviation for the clinic sample population. The MEI values are thus interpreted according to the following categories:

- MEI < 0.20 indicates that maintenance resources are utilised at high efficiency or are scarce, or both;
- $0.40 \geq MEI \geq 0.20$ reflects a reasonable range of maintenance efficiency, whereby the lower limit indicates good efficiency while the upper limit indicates low efficiency and/or slack of resources; and

- MEI > 0.40 indicates a high level of resources relative to the actual performance. Such high indicator values may express high maintenance expenditures, low physical performance because of ineffective maintenance, or a combination of these two extreme situations.

16.2.1.7 Maintenance sources diagram

Outsourcing constitutes an alternative to the implementation of maintenance activities by in-house employees, who require ongoing management. Outsourcing can serve as a source for the execution of seasonal preventive maintenance works, as well as rehabilitation, renovation and replacement works. This parameter reflects the mix of internal and external maintenance resources, and expresses the extent of outsourcing (in %) out of the total labour resources allocated for maintenance of the facility. Previous studies found that outsourcing may contribute to savings of about 10% compared with in-house provision (Domberger and Jensen, 1997). A mix that includes 60% outsourcing may provide a solid balance in healthcare facilities under standard service conditions.

16.3 Profile of clinic facilities

Table 16.2 depicts a profile of the clinic facilities sample, encompassing 42 data points. The mean floor area of the clinics is 1154 sq. m with an average age of 7.9 years. The respective mean AC for the population is 0.88, expressing the leanness of required maintenance resources for facilities with a mean age of 12.2 years. The annual number of patients per square metre, representing the clinic density, is 263.8, whereby an annual value of 175 is defined as standard. In light of the latter finding, we can deduce that the clinic facilities sample represents a facilities population that operates under intensive service conditions. The mean AME for the maintenance of the clinic sample population is $24.2 per sq. m, constituting an annual sum equal to 2.05% of the reinstatement value of the clinics ($1180 per sq. m). In light of the low average age of the clinic sample, this level of maintenance is high and can be explained by the intensive service conditions of the clinic facilities, as discussed in the preceding text. The MSD expresses the mixture of outsourcing and in-house maintenance service provision. The MSD shows that 45.5% of the services are contracted out. This mix is rationalised by the intensive service conditions, which require high availability of maintenance crews for urgent service that is provided by in-house maintenance crews. The managerial span of control (MSC) of the clinics' regional facility manager is 6.1 compared with a standard value of 6–8. The BPI's mean value is 95.5, indicating the clinic facilities' high performance. The parameter's relatively small variance indicates the significance of this parameter as a result of a performance-based maintenance policy. The last parameter – maintenance efficiency indicator (MEI) expresses the

Table 16.2 Summary of parameters for clinic facilities sample.

Parameter	N	Mean	Standard deviation
Floor area (sq. m)	42	1.154	1.148
Age (years)	42	7.9	6.1
Annual number of patients/sq. m	42	258.0	124.0
DC	42	1.16	0.15
Annual maintenance expenditure ($/sq. m)	42	24.2	13.7
Maintenance sources diagram (MSD)	42	45.34%	3.23%
Managerial span of control (MSC)	42	6.1	1.9
BPI	42	95.5	1.9
MEI	42	0.30	0.16

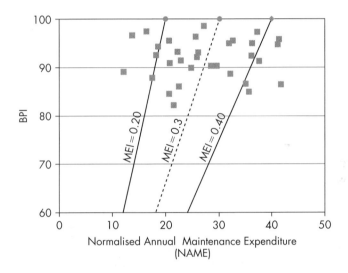

Figure 16.5 Building performance indicator versus normalised annual maintenance expenditure for clinic sample.

efficiency with which maintenance resources are utilised. The mean MEI equals the predicted normative analytical value (0.30).

Figure 16.5 depicts the distribution of the clinic facilities sample in a two-dimensional space, where the independent variable is the NAME and the dependent variable is the BPI. The three lines represent *lines of equivalent efficiency*, where the 0.30 line is the normative, and the other two express the margins of the standard region of efficiency. The distribution shows that 2/3 of the sample population falls between the upper and lower limits of the normative region. This distribution validates the predicted values deduced from the analytical development of this parameter. Facilities that are found close to the left margin (MEI = 0.20) exemplify high efficiency in implementing maintenance resources, whereas facilities found close to, or beyond the right MEI=0.40 line require further analysis to reveal the sources of inefficiency and consideration of an improvement programme.

16.3.1 Case study
The clinic under study is located in the centre of Tel Aviv. The built floor area of the clinic is 2180 sq. m. A detailed analysis of the clinic's KPIs and setting of a corrective policy were carried out in the framework of a 3-year three-stage case study: A clinic performance survey and an analysis of KPIs were executed in early 2003, referring to the financial year 2002 (Stage A). The need to implement a corrective maintenance policy was deduced and such was subsequently conducted in 2003–2004 along with a clinic performance and KPI analysis, completed in 2004 to examine the effectiveness and efficiency of the measures implemented on the strength of the KPI analysis (Stage B). The effectiveness of the corrective policy was validated by a KPI survey carried out at the end of the second year (Stage C). This survey revealed that the corrective maintenance policy improved performance of the exterior envelope and interior finishing systems, which contributed to the clinic's overall performance.

16.3.1.1 Stage A: KPI analysis – benchmarking year (2002)
Table 16.3 delineates the clinic's principal KPIs. A detailed KPI analysis revealed good performance ratings for clinic systems (BPI = 87.0), though the effectiveness and efficiency of activities were found to be poor, as the MEI level was 0.514, significantly higher than the upper margin of the normative region. This finding was attributed to high in-house and outsourcing costs.

Table 16.3 Clinic facility case study – 2002 versus 2004.

Parameter	2002	2004
AME	27.3	26.4
Number of patients	179	363
DC	1.01	1.31
MSD	73.3	64.8
MSC	9	2
BPI	87.0	91.2
MEI	0.514	0.296

16.3.1.2 Stage B: corrective policy implementation (2003–2004)

The corrective policy implemented during the years 2003–2004 included measures to increase managerial effectiveness by reducing the MSC at the regional FM level from 9 to 2, and increasing the DC by adding new functions, such as pharmacy and unification of clinics, that increased the number of insurees and patients. As a result, the DC increased from 1.01 to 1.31.

16.3.1.3 Stage C: KPI analysis – validation year (2004)

The performance survey conducted in 2005 which referred to 2004 data, found improved performance by the exterior envelope and interior finishing systems, which resulted in an improvement in overall performance, to a BPI level of 91.2. The AME was reduced during the same period from $27.3 to $26.4 per sq. m, primarily because of a decrease in outsourcing costs. The annual number of patients per square metre increased from 179 to 363. The integration of corrective measures – managerial, technical and business – resulted in a reduction of the MEI from 0.514 to 0.296, which is fairly close to the normative line.

16.4 Hospital facilities versus clinic facilities – comparative perspective

A concise FM view can be obtained by comparing several key parameters of clinic and hospital facilities: Table 16.4 presents such a comparison – it shows that while capital investments in clinics are two-thirds of the investments in hospital facilities, clinics exhibit a much higher performance – the performance of clinics is close to its designated level whereas hospital facilities perform at a marginal, yet satisfactory level. This can be seen clearly by comparing the efficiency and performance graphs of clinic facilities with those of hospital facilities (Figures 16.5 and 16.6, respectively). Figure 16.6 depicts the population of hospital facilities (Shohet et al., 2003); the normative range of MEI is 0.37–0.52, illustrating a lack of maintenance resources as well as poor performance of the built facilities. Two surveys of hospital facilities conducted within a period of 4 years revealed poor, but improving performance: the BPI

Table 16.4 Key parameters of clinic versus hospital facilities.

Parameter	Hospitals	Clinics
Reinstatement value ($/sq. m)	1780	1180
AME ($/sq. m)	54.5	29.5
BPI	76.6	95.5
Built floor area per insuree	0.30	0.15
MEI	0.44	0.30

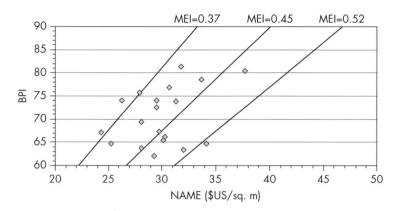

Figure 16.6 Building performance indicator (BPI) versus normalised annual maintenance expenditure (NAME) for hospital buildings.

of a sample of 17 hospitals indicated a deteriorating level (BPI=68.8) (Shohet *et al.*, 2003), and a survey conducted later, in 2004 showed improvement of the BPI, to 76.6 (Shohet and Lavy, 2004). Both samples show that built facilities of hospitals perform considerably poorer than clinic facilities because of scarcity of resources. The required investments in maintenance of hospital facilities are more than 60% higher, whereas the required built floor area per insuree is about half of that of clinics (0.15 compared with 0.30). This comparison sheds light on the potential improvements in healthcare built environments that can be attained from the development of clinics for delivery of close healthcare.

16.5 Concluding remarks

Comparing hospital facilities with clinics reveals a considerable difference between the facilities in terms of FM parameters: while hospital facilities necessitate the allocation of double the amount of resources for maintenance, it is hard to attain high performance in hospital built facilities because of the size of the facility and complexity of the electro-mechanical systems which pose technological barriers to accomplishing their designated performance. Furthermore, the built floor area per insuree is significantly lower in clinic facilities (0.15 sq. m) than in acute-care hospitals (0.30 sq. m), thus the actual expenditure per insuree in hospital facilities is four times higher than in clinic facilities.

Community clinics are much accessible than peripheral or regional hospitals in terms of geographical dispersion, a greater scope of close care may be provided, and the annual cost of facilities per insuree is approximately 75% lower than in regional hospitals. From a life cycle perspective, maintenance of hospital facilities is characterised by multiple maxima and minima, that reflect the replacement of electro-mechanical systems.

Provision of some hospital functions (outpatient clinics, complimentary care and rehabilitation) in clinic facilities instead of peripheral and regional hospital facilities establishes a potential for flexible infrastructures that can be maintained at high performance with a lesser extent of resources. Furthermore, such architecture may reduce the size of the core hospital and contribute to a reduction in the complexity of maintenance of the core hospital facility. It is anticipated that the development of community clinic infrastructures will replace a significant share of the existing infrastructures in hospitals. The development of clinic infrastructures entails improving the provision of built facilities for healthcare and improving the accessibility to close care of populations such as the elderly and families with children. Such facilities will result in savings in both capital investment in built facilities and AME, and will contribute to improvement in the performance of healthcare built facilities.

References

American Hospital Association (AHA). (2006). "TrendWatch Chartbook 2004: Trends affecting hospitals and health systems – September 2004", available at http://www.hospitalconnect.com/ahapolicyforum/trendwatch/chartbook2006.html, September 2007, Chapter 4, p. 2.

British Ministry of Finance. (2003). "Budget 2003: Report – Chapter 6: Delivering high quality public services", available at http://www.hm-treasury.gov.uk/budget/bud_bud03/budget_ report/bud_bud03_repchap6.cfm, February 2005.

Domberger, S. and Jensen, P. (1997). "Contracting out by the public sector: Theory, evidence, prospects", *Oxford Review of Economic Policy*, 13(4), 67–78.

Federal Statistical Office Germany. (2003). "Facilities, beds and patient turnout: Hospitals, 1991–2003", available at http://www.destatis.de/basis/e/gesu/gesutab29.htm, June 2004.

Hensher, M. and Edwards, N. (1999). "The hospital of the future: Hospital provision, activity, and productivity in England since 1980", *British Medical Journal*, 319(7214), 911–914.

Hosking, J. E. and Jarvis, R. J. (2003). "Developing a replacement facility strategy: Lessons from the healthcare sector", *Journal of Facilities Management*, 2(3), 214–228.

Israel Standards Institution. (2002). "*IS-1525: Part 1 – Building maintenance management: Elements and finish*", Israel Standards Institution, Tel-Aviv, Israel (in Hebrew).

Melin, A. and Granath, J. A. (2004). "Patient focused healthcare: An important concept for provision and management of space and services to the healthcare sector", *Facilities*, 22(11/12), 284–289.

Rees, D. (1997). "The current state of facilities management in the UK National Health Service: An overview of management structures", *Facilities*, 15(3/4), 62–65.

Rees, D. (1998). "Management structures of facilities management in the National Health Service in England: A review of trends 1995–1997", *Facilities*, 16(9/10), 254–261.

Shohet, I. M. and Kot N. (2006). "Examination of parameters for maintenance of clinics facilities", Proceedings of the CIB W70 Changing User Demands on Buildings: Needs for Lifecycle Planning and Management, 12–14 June, 2006, Trondheim, Norway, ISBN 82-7551-031-7.

Shohet, I. M., Kot, N., and Karako, I. (2008). "Clinics facilities maintenance using life cycle costs principles", Proceedings of the CIB W-70 International Conference in Facilities Management, Heriot Watt University, Edinburgh, 16th–18th June, 2008, pp. 259–268.

Shohet, I. M. and Lavy, S. (2004). "Development of an integrated healthcare facilities management model", *Facilities*, 22(5/6), 129–140.

Shohet, I. M., Lavy-Leibovich, S., and Bar-On, D. (2003). "Integrated maintenance management of hospital buildings", *Construction Management of Economics*, 21(2), 219–228.

Shohet, I. M., Leibovich-Lavy S., and Bar-On, D. (2002). "Integrated maintenance management of hospital buildings in Israel", Proceedings of the 17th International Symposium of the International Federation of Hospital Engineering, Bergen, 12–16 May, 2002.

World Bank. (2007). "World Bank series", available at http://devdata.worldbank.org/query, September 2007.

World Healthcare Organization. (1978). Alma Ata 1978: Primary Health Care, HFA Sr. No. 1.

Index